The ASPHALT BINDER HANDBOOK

ASPHALT INSTITUTE

MS-26
1st Edition

ASPHALT INSTITUTE
MANUAL SERIES NO. 26 (MS-26)

The Asphalt Institute can accept no responsibility for the inappropriate use of this manual. Engineering judgment and experience must be used to properly utilize the principles and guidelines contained in this manual, taking into account available equipment, local materials, and conditions.

All reasonable care has been taken in the preparation of this manual; however, the Asphalt Institute can accept no responsibility for the consequences of any inaccuracies it may contain.

Printed in USA
First Edition
ISBN: 978-1-934154-63-2
Library of Congress Control Number: 2011906807
Copyright © 2011
All Rights Reserved
Asphalt Institute

Foreword

In 1995, *SP-1, Superpave: Performance Graded Asphalt Binder Specification and Testing*, was published describing the tests and specification recommended by the Strategic Highway Research Program. The manual was unique in that it was the first major Asphalt Institute publication devoted exclusively to the subject of asphalt binders.

Twelve years later, in 2007, *MS-25, Asphalt Binder Testing: Technician's Manual for Specification Testing of Asphalt Binders*, was first published, taking the subject of testing of asphalt binders to a new level. Very detailed descriptions of test procedures were provided to allow the technician an in-depth examination of performance graded (PG) asphalt binder testing. MS-25 is updated frequently to keep pace with ASTM and AASHTO test procedure changes and is a cornerstone text used in the Institute's National Binder Technician Certification program as well as a reciprocal program offered at NETTCP in New England.

Also in 2007, the 7th Edition of *MS-4, The Asphalt Handbook*, was published as a major update to this venerable text. For more than 70 years *The Asphalt Handbook* has served as the Asphalt Institute's comprehensive manual on the use of asphalt. It incorporates information found in other AI publications such as MS-1, MS-2, and MS-22—serving the industry as the single source on asphalt.

Now the Asphalt Institute is pleased to offer its latest publication, *MS-26, The Asphalt Binder Handbook*, designed to offer the industry a single-source reference on asphalt binders. This new manual features new information never before published in an AI manual while incorporating some information found in other manuals, such as SP-1 and MS-25. This is the Asphalt Institute's "little black book" of all things asphalt.

Acknowledgments

In some aspects, MS-26 is a compendium of information from other manuals and documents, and thus it is virtually impossible to thank all of the contributors to its content and publication. However, the efforts of a few people and groups are certainly noteworthy.

Thanks are due in particular to Bob McGennis (Holly Asphalt, formerly Asphalt Institute), Dr. Hussain Bahia (University of Wisconsin-Madison, formerly Penn State University), Dr. Scott Shuler (Colorado State University, formerly Asphalt Institute) and John Bukowski (Federal Highway Administration) for their effort in writing/reviewing FHWA Report SA-94-069, which became the basis for SP-1. Thanks are also due to Dr. Dave Anderson (Consultant) and Mike Beavin (Asphalt Institute) for their efforts in writing MS-25 and their work is also reflected here.

A critical evaluation is also necessary to ensure that a publication is technically correct. For MS-26, thanks are expressed to Shauna TecleMariam (US Oil & Refining), Don Powell (San Joaquin Refining), Dr. John D'Angelo (Consultant), and the engineering staff of the Asphalt Institute—particularly Danny Gierhart, Greg Harder, and Bob Humer—for their thorough reviews.

Finally, some of the elements in this publication come from technical content contributed by the Technical Advisory Committee of the Asphalt Institute. Thanks are due to this committee, its leadership and members, for their efforts in contributing to the technical content herein.

As always, we express our gratitude to the Members of the Asphalt Institute. Without their interest and support, this publication would not exist.

R. Michael Anderson
Asphalt Institute
March 2011

Table of Contents

Foreword iii
Acknowledgments v
Table of Contents vii
List of Illustrations xiii
List of Tables xvii

CHAPTER 1   INTRODUCTION **1**
    What is asphalt? 3
    Natural asphalts 4
    Petroleum asphalts 6
        Refining crude petroleum 6
        Crude selection 8
    Asphalt usage 10

CHAPTER 2   ASPHALT CHEMISTRY **13**
    Chemical composition 15
        Types and structure of functional groups 15
            Polarity 17
            Oxidation 17
            PAHs 19

CHAPTER 3   ASPHALT PHYSICAL PROPERTIES AND CHARACTERISTICS **21**
    Physical properties 23
        Consistency 23
        Purity 23
        Safety 24
    How asphalt behaves 25
        Fundamental vs. empirical properties 30
    Sampling, heating and splitting asphalt binders . . . 31
        Sampling 31
        Heating and splitting asphalt binder samples . . . 33
        Changes in behavior with storage
        and handling 35

CHAPTER 4   ASPHALT CEMENTS **37**
    Historical tests 39
        Viscosity 39
            Absolute viscosity 39
            Kinematic viscosity 41
            Apparent viscosity 43
        Penetration 45
        Flash point 45
        Aging 49

Thin-Film Oven Test (TFOT) 50
Rolling Thin-Film Oven Test (RTFOT). 50
Ductility 53
Solubility 56
Specific gravity 58
Specifications for asphalt cements 59
   Penetration-Graded specification 59
      Advantages and shortcomings 59
   Viscosity-graded specification 60
      The AC system. 60
      The AR system. 61
      Advantages and shortcomings 61
PG (Performance Graded) asphalt
   binder tests 64
   Rotational viscometer 64
   Dynamic Shear Rheometer (DSR) 67
   Bending Beam Rheometer (BBR). 77
   Direct Tension Tester (DTT) 84
   Aging procedures 91
PG asphalt binder specification 95
   Advantages compared with older
      specifications. 97
   Assumptions in the specification 98
   How the PG specification addresses
      pavement performance 99
      Safety 99
      Pumping and handling 99
      Permanent deformation 100
      Excessive aging 101
      Long-term stiffness 101
      Thermal cracking. 101
   Grading an asphalt binder in the
      PG specification system 104
      Non-grade determination requirements 105
      Verifying asphalt binder Performance
         Grade using AASHTO M 320 Table 1. 105
      Grading an unknown sample. 106
      Approximations/rules of thumb 106
      Critical temperature, continuous grade,
         and UTI 107
      Asphalt binder grade selection. 109
      Grade bumping for greater traffic loading. . . . 114
   Performance-Graded asphalt binder using
      the Multiple Stress Creep Recovery
      (MSCR) test 115
      MSCR test and specification 116

CHAPTER 5 ASPHALT EMULSIONS AND CUTBACKS ... 119

Emulsified asphalts: history and uses. ... 121
Composition. ... 122
Types and classifications. ... 122
Emulsion components and production. ... 124
    Asphalt. ... 124
    Water ... 125
    Emulsifying agents. ... 125
    Production ... 126
Breaking and curing. ... 128
    Factors affecting breaking and curing. ... 129
Variables affecting emulsion quality. ... 129
Storing asphalt emulsions. ... 131
Handling asphalt emulsions ... 132
Sampling asphalt emulsions ... 133
Tests for emulsified asphalts. ... 135
    Saybolt Furol viscosity. ... 135
    Storage stability and settlement. ... 136
    Classification. ... 137
    Demulsibility ... 137
    Cement Mixing. ... 137
    Coating ability and water resistance. ... 138
    Particle charge ... 138
    CSS classification (mixing). ... 138
    Sieve. ... 139
    Density. ... 139
    Breaking Index. ... 139
    Distillation ... 140
    Tests on residue ... 141
        Float test. ... 141
Specifications for emulsified asphalts ... 143
Cutbacks: Introduction. ... 145
    Uses of cutback asphalts ... 145
Types and classifications of cutback asphalts ... 145
Tests for cutback asphalts. ... 147
    Kinematic viscosity. ... 147
    Tag Open Cup Flash Point. ... 147
    Water. ... 147
    Distillation ... 147
    Tests on residue ... 148
        Residue of 100 penetration ... 149
Specifications for cutback asphalts. ... 150

CHAPTER 6	AIR-BLOWN ASPHALT	151
	Roofing	153
	Properties, specification, and tests	154
	Tests	154
	Ring-and-Ball Softening Point	154
	Penetration	157
	Loss on heating	157
	Rotational viscosity	158
	Specifications	158
CHAPTER 7	ASPHALT MODIFICATION	161
	Types of modifiers and additives	163
	Elastic modification (elastomers)	164
	Plastic modification (plastomers)	166
	Crumb rubber modification	166
	Mixed modification	167
	Chemical modification	167
	Extenders	168
	Oxidants and antioxidants	168
	Hydrocarbons	168
	Antistripping additives	169
	Additional tests for modified asphalt binders	169
	Separation tests—ointment tube	169
	Solubility—centrifuge	170
	Recovery and stress-strain tests	170
	Elastic Recovery	171
	Force Ductility	171
	Toughness and tenacity	174
	DSR phase angle	174
	DSR Creep-Recovery	176
CHAPTER 8	SAFETY	179
	Safe handling of asphalt	181
	Storage temperatures	181
	Asphalt fume	182
CHAPTER 9	UNDERSTANDING TESTING VARIABILITY	183
	Repeatability and reproducibility	185
	Accuracy, precision, and bias	185
	Variability in asphalt binder tests	187
	Dispute resolution	192
CHAPTER 10	SUPPLEMENTAL TOPICS	197
	Use of solubility in the specifications	199
	Asphalt rheology: mastercurves	201

Mixing and compaction temperatures
for asphalt binders 205
Mixing temperatures 205
HMA and WMA 206
Determining laboratory mixing and
compaction temperatures 207
Viscosity procedure 207
Phase angle procedure 211
Steady Shear Flow (SSF) procedure 211
Temperature-Volume relationships 214
Density and specific gravity 214
Temperature-volume relationship
and calculations 214
Specific gravity calculations 215
Tank measurements 216
Commonly-asked miscellaneous
questions 218
What is the thermal conductivity of
asphalt binder? 218
What is the vapor pressure of asphalt
at typical storage temperatures? 218
What is the typical thermal BTU value
for a pound of asphalt? 218
What is the typical value for the specific
heat of asphalt? 218

INDEX 219

List of Illustrations

Chapter 1
Figure 1.1	Trinidad Lake asphalt	4
Figure 1.2	TLA composition and particle size distribution	5
Figure 1.3	Rancho LaBrea	5
Figure 1.4	Petroleum asphalt production	8
Figure 1.5	Distillation products	9
Figure 1.6	Global bitumen use and application	10
Figure 1.7	U.S. Asphalt usage by year	10

Chapter 2
Figure 2.1	SEC of asphalts	18
Figure 2.2	Effect of aging on stiffness	19

Chapter 3
Figure 3.1	Time-temperature superposition	25
Figure 3.2	Behavior of asphalt binder as shear rate changes	26
Figure 3.3	Asphalt binder stiffness as a function of temperature	27
Figure 3.4	Effect of temperature on property change	28
Figure 3.5	Effect of physical hardening	30
Figure 3.6	Sample filling	33
Figure 3.7	Heating a sample on a hot plate	34
Figure 3.8	Splitting a sample	35

Chapter 4
Figure 4.1	AIVV Tube	40
Figure 4.2	Absolute viscosity bath with tubes	40
Figure 4.3	Pouring absolute viscosity tubes	40
Figure 4.4	Cross-arm viscosity tube	41
Figure 4.5	Kinematic viscosity bath with tubes	42
Figure 4.6	Apparent viscosity graph	44
Figure 4.7	Penetrometer	46
Figure 4.8	Penetration test	47
Figure 4.9	COC test	47
Figure 4.10	COC overview (manual)	48
Figure 4.11	COC manual operation	48
Figure 4.12	COC flash	49
Figure 4.13	TFO	51
Figure 4.14	RTFO bottles	51
Figure 4.15	Bottle rotation	52
Figure 4.16	Cooling rack (ASTM)	52
Figure 4.17	RTFO	52
Figure 4.18	Draining RTFO residue into container	53

Figure 4.19	Scraping tool	54
Figure 4.20	Scraping RTFO residue into container	54
Figure 4.21	Mold preparation and pouring	55
Figure 4.22	Trimming ductility specimen	55
Figure 4.23	Ductility test	56
Figure 4.24	Solubility test	57
Figure 4.25	Solubility crucibles	58
Figure 4.26	Specific gravity pycnometers	58
Figure 4.27	Penetration grading–asphalt binder properties	60
Figure 4.28	Viscosity grading–asphalt binder properties	62
Figure 4.29	Pen-Vis correlation for MRL asphalt binders	63
Figure 4.30	RV	65
Figure 4.31	RV operation	66
Figure 4.32	Inserting sample chamber into thermal chamber	66
Figure 4.33	Graphic representation of shear modulus and phase angle	68
Figure 4.34	DSR operation	69
Figure 4.35	DSR	69
Figure 4.36	Direct pour	70
Figure 4.37	Silicone mold	71
Figure 4.38	Transfer to top plate	71
Figure 4.39	Transfer to bottom plate	72
Figure 4.40	Trimming	73
Figure 4.41	Asphalt sample configuration	73
Figure 4.42	Sample after trimming	74
Figure 4.43	Stress-strain output	75
Figure 4.44	Stress-strain output of VE material	75
Figure 4.45	DSR calculations	76
Figure 4.46	Sample report	76
Figure 4.46	Sample report	77
Figure 4.47	BBR	78
Figure 4.48	BBR operation	79
Figure 4.49	BBR mold	79
Figure 4.50	Pouring	80
Figure 4.51	Properly poured test specimen	80
Figure 4.52	Trimming	81
Figure 4.53	Graphs	82
Figure 4.54	Sample report	83
Figure 4.55	DTT	84
Figure 4.56	DTT loading	85
Figure 4.57	DTT specimen	86
Figure 4.58	DTT stress curves	87
Figure 4.59	Mold assembly	87
Figure 4.60	Pouring	88
Figure 4.61	DT specimens poured	88
Figure 4.62	Trimming	89

Figure 4.63	Loading	89
Figure 4.64	Types of breaks	90
Figure 4.65	DTT stress-strain curves	90
Figure 4.66	PAV	92
Figure 4.67	PAV pans	93
Figure 4.68	Scraping	93
Figure 4.69	PG binder tests with aging	95
Figure 4.70	PG specifications table	96
Figure 4.71	Critical cracking temperature	97
Figure 4.72	PG binder tests with temperature	98
Figure 4.73	Rutting	100
Figure 4.74	Fatigue cracking	102
Figure 4.75	Thermal cracking – early	103
Figure 4.76	Thermal cracking – late	103
Figure 4.77	Design temperature reliability	111
Figure 4.78	High and low air temperature variations	112
Figure 4.79	High and low pavement temperature variations	112
Figure 4.80	PG asphalt binder grade selections	113
Figure 4.81	Creep-recovery response from MSCR	116
Figure 4.82	Non-recoverable strain	117
Figure 4.83	Calculating J_{nr}	117
Figure 4.84	Calculating MSCR recovery	118

Chapter 5

Figure 5.1	Schematic of emulsifying plant	127
Figure 5.2	Asphalt particles in emulsion	128
Figure 5.3	Saybolt Furol viscosity	136
Figure 5.4	Emulsion Distillation	140
Figure 5.5	Float test	142
Figure 5.6	Cutback distillation	148

Chapter 6

Figure 6.1	Filling R&B molds	155
Figure 6.2	Trimming R&B specimens	155
Figure 6.3	SP apparatus (manual)	156
Figure 6.4	Ring and Ball Softening Point (automatic)	157

Chapter 7

Figure 7.1	SBR Latex Modification of Hot Asphalt Cement	164
Figure 7.2	SBS Pellets	165
Figure 7.3	SBS blending system for Asphalt Terminals	165
Figure 7.4	Ointment tubes	170
Figure 7.5	ER test specimen	172
Figure 7.6	ER test operation	172
Figure 7.7	ER specimens cut	173
Figure 7.8	ER specimens recover	173
Figure 7.9	FD test operation	174
Figure 7.10	FD output	175

Figure 7.11	T&T equipment and operation	175
Figure 7.12	T&T output	176
Figure 7.13	Calculating MSCR recovery	176
Figure 7.14	Determining elastic behavior from MSCR	177

Chapter 9

| Figure 9.1 | Dispute resolution flow chart | 195 |

Chapter 10

Figure 10.1	Mastercurve sample data	202
Figure 10.2	Mastercurve isotherms	203
Figure 10.3	Mastercurve G^* and phase angle	203
Figure 10.4	Mastercurve parameters	204
Figure 10.5	Rheological properties of different angles	204
Figure 10.6	Temperature-viscosity example for mixing and compaction temperatures	208
Figure 10.7	Temperature-viscosity graph	210
Figure 10.8	Mixing and compaction temperature determination using phase angle procedure	212
Figure 10.9	Comparison of viscosity and phase angle procedures	213
Figure 10.10	Comparison of viscosity and Steady Shear Flow (SSF) procedures	213

List of Tables

Chapter 1
Table 1.1	Methods used to produce and process asphalt materials	7
Table 1.2	Typical residuum content of selected crude oils	9

Chapter 2
Table 2.1	Elemental analysis of asphalts from various sources	15

Chapter 4
Table 4.1	Asphalt cement specification tests	63
Table 4.2	Comparison of AASHTO M320 grade with continuous grade	109

Chapter 5
Table 5.1	Major uses of asphalt emulsions	121
Table 5.2	Standard emulsified asphalt grades	124
Table 5.3	Asphalt storage guidelines	131
Table 5.4	Storage temperatures	131
Table 5.5	Handling asphalt emulsions	132
Table 5.6	Specification tests for emulsified asphalt grades	144
Table 5.7	Specification tests for cutback asphalt grades	150

Chapter 6
Table 6.1	Air-blown asphalt specification tests (ASTM)	159

Chapter 9
Table 9.1	Precision statement for AASHTO T 313 – BBR	187
Table 9.2	Asphalt binder test variability (1s percent and d2s percent)	190
Table 9.3	Asphalt binder test variability (1s and d2s)	191

Chapter 10
Table 10.1	Ranges of storage and mixing temperatures	206
Table 10.2	Percent capacities for various depths of cylindrical tanks in horizontal position	217

CHAPTER 1

Introduction

- **What is asphalt?**
- **Natural asphalts**
- **Petroleum asphalts**
    Refining crude petroleum
    Crude selection
- **Asphalt usage**

What is asphalt?

According to the American Society for Testing and Materials (ASTM), asphalt is a dark brown to black cementitious material. Its predominating constituents are bitumens, which occur in nature or are obtained in petroleum processing.

The reader should note that the terms *bitumen* and *asphaltic bitumen* are used in Europe and are synonymous with the term *asphalt* used in North America. Outside North America, the term *asphalt* is used to describe mixtures of bitumen with mineral aggregates. To avoid confusion, in North America liquid asphalt is often also referred to as "asphalt cement" or, more generically, as "asphalt binder."

Asphalts are viscous liquids or solids essentially consisting of hydrocarbons and their derivatives, which are soluble in carbon disulfide (CS_2). They are substantially nonvolatile at ambient temperatures and soften gradually when heated. Asphalts are often referred to as viscoelastic materials, behaving as elastic solids at low temperatures and viscous liquids at high temperatures.

Asphalt is present in varying proportions in most crude petroleum sources. Although asphalts also occur as natural asphalts, rock asphalts, and lake asphalts, the majority of asphalt material used today is obtained from petroleum through various refining processes.

Asphalts should not be confused with coal-tar products such as coal tar or coal-tar pitches. The latter are manufactured by the high-temperature carbonization of bituminous coals and differ from asphalts substantially in composition and physical characteristics. The differences between asphalts and coal-tar products are well defined in the literature.

Similarly, asphalts should not be confused with petroleum pitches, which are often highly aromatic residues, produced by thermal cracking, coking, or oxidation from selected petroleum fractions. Petroleum pitches are primarily used as binders in the manufacture of metallurgical electrodes.

Natural asphalts

Natural asphalts are, as implied by the name, naturally-occurring asphalts that are not refined from petroleum. Lake asphalts, like Trinidad Lake asphalt (TLA) on the island of Trinidad (see Figure 1.1), result from the evaporation of volatile portions of natural deposits, leaving asphalt fractions. This evaporation also leaves some insoluble material in the asphalt fraction. The composition and particle size distribution of TLA is shown in Figure 1.2.

As can be seen in Figure 1.2, although mineral matter constitutes a substantial percentage of a TLA sample, the majority of the mineral matter present would pass through a 0.038-millimeter (#400) sieve. Nevertheless, this mineral matter contributes to a stiffening of the TLA.

Other natural asphalts can be found in "lake" form including Bermudez in Venezuela and the La Brea Tar Pits in California (Figure 1.3). Additional, natural asphalts may also be found penetrating aggregate. Examples of this type of natural asphalt include rock asphalt in Kentucky and the Athabasca tar sands in Canada. Despite the use of the word "tar," these deposits contain asphalt and not tar.

Figure 1.1 Trinidad Lake asphalt

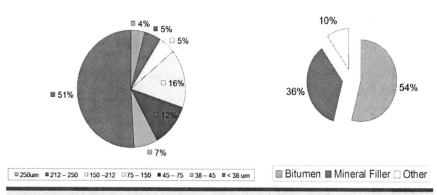

Figure 1.2 TLA composition and particle size distribution

Figure 1.3 Rancho LaBrea

Although natural asphalts were commonly used early in the 20th century, their use—relative to total asphalt usage—decreased as petroleum asphalts became more common. Today, natural asphalts are rarely used without being blended with petroleum-derived asphalts. In some cases, natural asphalts are considered to be a modifier to standard petroleum asphalts.

Petroleum asphalts

Most asphalt produced in the United States is the result of the crude oil distillation process. Even though vacuum distillation is the basic unit of operation, additional manufacturing methods can also be used. The choice of manufacturing methods depends on the properties of crude available. Because crude oil chemical composition varies significantly, the properties and yield of residuum (the base material for the production of asphalt) can also vary over a wide range.

Refining crude petroleum

As the API (American Petrochemical Institute) gravity of crude oil increases, the yield of residuum decreases and the yield of distillate subsequently increases. The industry defines crude oil as "heavy" when its API gravity is less than about 25. These low API gravity crude oils are also called "sour" crude oils if their sulfur content is high. If a crude oil has an API gravity greater than about 25 and is generally low in sulfur content, it is called a "light" and "sweet" crude oil. Most fully integrated refineries are designed to maximize distillate products and are not generally adapted to process heavy crude oils, possibly because of the investment required in processes to extract the sulfur. This limitation can restrict the refiners' options for crude selection and ability to produce asphalts. On the other hand, asphalt refineries are designed to maximize the production of heavy petroleum products and the refiner intentionally selects heavy crude oils for processing.

Several manufacturing methods are available to produce specification asphalts depending on the crude source(s) and processing capabilities available. Often a combination of processes is selected. Table 1.1 provides a summary of the manufacturing methods that may be used by an asphalt producer.

The most common practice for producing asphalt cements is straight reduction to grade from crude oil or a crude blend using atmospheric and vacuum distillation. The petroleum asphalt flow chart (Figure 1.4) shows, in general, the flow of crude petroleum through a refinery. The emphasis is on that part of the process related to the refining and production of asphalt cement.

Atmospheric distillation is used to separate various "overhead" crude fractions and a non-boiling component known as an atmospheric residue. Crude flows continuously through a furnace that heats the oil and strips out any light materials present. The heated crude flows through an atmospheric tower where light components flash to the upper reaches

Table 1.1
Methods used to produce and process asphalt materials

Production/Process	Base Materials	Product
Atmospheric and Vacuum Distillation	Asphalt-based crude or crude mix	Asphalt cements
Blending	Hard and soft asphalt	Asphalt cements of intermediate consistency
	Asphalt cements and petroleum distillates	Cutback asphalts
Air-Blowing	Asphalt flux	Asphalt cements, roofing asphalts, pipe coatings, special membranes
Solvent Deasphalting	Vacuum residuum	Hard asphalt
Solvent Extraction	Vacuum residuum	Asphalt components (asphaltenes, resins, oils)
Emulsification	Asphalt, emulsifying agents, and water	Emulsified asphalts
Modification	Asphalt and modifiers (polymers, chemicals, etc.)	Modified asphalts

of the tower and subsequent lower boiling components reach lower sections of the tower. Different boiling components are collected as they condense and are removed for further processing (Figure 1.5). The overhead fractions include gasoline (60°F–325°F), naphtha (300°F–400°F), kerosene (325°F–500°F), light gas oil (450°F–600°F), and heavy gas oil (575°F–700°F).

The non-boiling atmospheric residue is introduced into a vacuum distillation unit in which several high boiling distillate fractions are produced, leaving a vacuum residue. The overhead fractions are gas oils (650°F–750°F and 700°F–850°F). The vacuum residue is used as the base material for the production of asphalt. The atmospheric equivalent vapor temperature to yield the vacuum residue is about 1000°F.

The vacuum residue is also the feed stock of several operations that are designed to increase the yield of refined products. Some of these processes yield hard asphalts that can be used as blends with other components, such as atmospheric and vacuum residua, to produce asphalts.

The processes that produce hard asphalt components for blending include solvent extraction (ROSE), solvent deasphalting (SDA) or propane deasphalting (PDA), and air blowing or blowing rectification (OX). Practically, refineries that produce asphalt cements may use all of these processes in combination and will produce several grades of asphalts, which are further blended to produce intermediate grades.

Air-blown asphalts are produced by blowing air at high temperatures through asphalt of the proper consistency. These asphalts are discussed later.

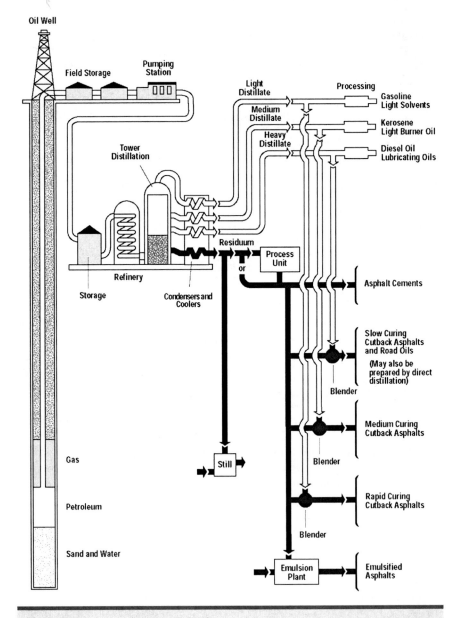

Figure 1.4 Petroleum asphalt production

Crude selection

As discussed in the previous section, the selection of a crude or crude blend determines the amount of residuum that can be recovered from the crude. The chemistry and composition of crude oil is highly dependent on its source. Geological age is an important factor in the consistency and

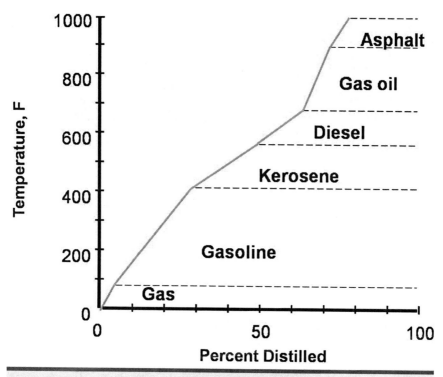

Figure 1.5 Distillation products

composition of crude oil, with older crude oils tending to be heavier (i.e., having a higher percentage of residuum). Table 1.2 shows some typical residuum contents for different crude oils.

Some crude oils may also contain large amounts of paraffinic wax that generally result in asphalts of poor quality. Wax contents in asphalt greater than 5 percent are usually considered undesirable.

Table 1.2
Typical residuum content of selected crude oils

Source	Residuum, %
California Valley (Kern River)	66
Boscan (Venezuela)	58
California Coastal (Hondo)	48
Alaska (North Slope)	31
Arabian Heavy	27
Nigeria Light	1

Asphalt usage

Approximately 100 million metric tonnes of asphalt are used worldwide on an annual basis. It is a very versatile material that has many industrial applications. More than 150 products are made with asphalt, but by far its greatest use is in road construction and maintenance. Worldwide, approximately 85 percent of all asphalt is used in paving applications. These include paving grade asphalt cements used for hot mix asphalt, emulsified asphalts

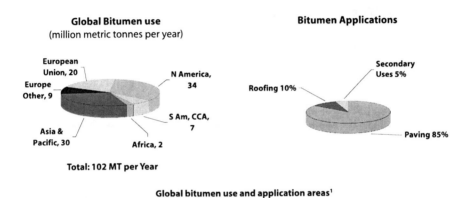

Figure 1.6 Global bitumen use and application

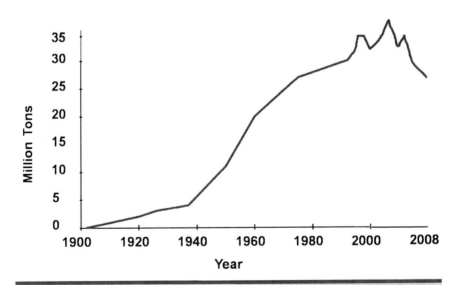

Figure 1.7 U.S. asphalt usage by year

for tack coats, cold mixtures, chip seals, slurry seals, and microsurfacing, or cutback asphalts for prime coats and cold mix. The remaining 15 percent is mostly comprised of asphalt for roofing applications, with some small percentage used in other industrial applications. Figure 1.6 shows asphalt usage by application and geography.

Currently, the United States uses approximately 22 million metric tonnes (2008 data) of asphalt annually in paving applications. Figure 1.7 illustrates the growth of asphalt usage in the United States since the start of the 20th century. As shown in the figure, asphalt usage was relatively low and growth steady from 1900–1940. From 1940–1980, the use of asphalt grew significantly. This growth coincided with the development of the interstate highway system and the addition of highway capacity to meet the needs of the transportation system. Since 1980, growth has been much slower as the focus has turned to maintenance, rehabilitation, and reconstruction rather than the addition of capacity through new construction.

CHAPTER 2

Asphalt chemistry

Chemical composition
Types and structure of functional groups
PAHs

Chemical composition

Asphalts can generally be described as complex mixtures containing a large number of different chemical compounds of relatively high molecular weight. The chemical composition of asphalts represents a wide range of organic molecules, with variation depending on the composition of the original crude oil and on the processes used during refining and blending.

There is considerable uncertainty as to the molecular weight distribution of asphalt. The smallest size, approximately 300, is determined by the cut point at the distillation of the asphalt. The largest size has not been conclusively determined, with some research suggesting that there are molecular weights up to 10,000, and other research indicating that there are few, if any, molecules larger than 1,000 in asphalt.

A typical elemental analysis is given in Table 2.1. Asphalts contain predominantly cyclic hydrocarbons (aromatic and/or naphthenic) and a lesser amount of saturated components which are mainly of very low chemical reactivity.

Some of these elements represent trace impurities originating from the particular crude used.

Types and structure of functional groups

The chemical characterization of asphalt is based on separation into four broad classes of compounds—asphaltenes, resins, cyclic, and saturates—using solvent precipitation and adsorption chromatography. The relative groups of compounds vary.

The molecules present in asphalt are combinations of alkanes, cycloalkanes, aromatics and heteromolecules containing sulfur, oxygen, nitrogen,

Table 2.1
Elemental analysis of asphalts from various sources

Element	Average	Range
Carbon, %w	82.8	80.2–84.3
Hydrogen, %w	10.2	9.8–10.8
Nitrogen, %w	0.7	0.2–1.2
Sulfur, %w	3.8	0.9–6.6
Oxygen, %w	0.7	0.4–1.0
Nickel, ppm	83	10–139
Vanadium, ppm	254	7–1590
Iron, ppm	67	5–147
Manganese, ppm	1.1	0.1–3.7
Calcium, ppm	118	1–335
Magnesium, ppm	26	1–134
Sodium, ppm	63	6–159

and heavy metals. The hetero-atoms, sulfur, nitrogen and oxygen, are typically found in the asphaltene and resin components.

Asphalt functionality relates to how molecules interact with each other and/or with aggregate surfaces and/or other materials. The content of nitrogen and oxygen in some molecules make them slightly polar, but the limited amount of nitrogen and oxygen prevent very polar molecules. The significance of molecules containing hetero-atoms in asphalt chemistry is the ability to form molecular associations, which strongly influence the physical properties and performance. The resin and asphaltene components containing the hetero-atom compounds vary in their characteristics in asphalts obtained from different crude sources. The predominant sulfur compounds are thiophenes. The predominant nitrogen compounds are based on pyrrole, indole, and carbazole groups. Predominant oxygen compounds appear to include furans, phenols, and carboxylic/naphthenic acid groups.

Asphaltenes are black amorphous solids containing carbon, hydrogen, nitrogen, sulfur, and oxygen. Trace elements such as nickel and vanadium are also present. Asphaltenes are generally considered as highly polar aromatic materials of high molecular weight. They typically constitute 5–25 percent of the weight of asphalt.

Saturates comprise predominantly the straight and branched-chain aliphatic hydrocarbons present in asphalt, together with alkyl naphthenes and some alkyl aromatics. The average molecular weight range is similar to that of the cyclic. The components include both waxy and non-waxy saturates. This fraction typically constitutes 5–20 percent of the weight of asphalt.

Cyclics (naphthene aromatics) comprise the compounds of lowest molecular weight and represent the major portion of the dispersion medium for the peptized asphaltenes. They are dark, viscous liquids that typically constitute 45–60 percent of the weight of the total asphalt. Cyclics are compounds that have aromatic and naphthenic aromatic nuclei with side chain constituents. They have molecular weight in the range of 500–900.

Resins (polar aromatics) are dark-colored, solid or semi-solid, very adhesive fractions of relatively high molecular weight present in the maltenes. They are dispersing agents (referred to as peptizers) for the asphaltenes. The proportion of resins to asphaltenes governs, to a degree, the rheological behavior of asphalt. Resins have a molecular weight in the range of 800–2000, but with a wide molecular distribution. They typically constitute 15–25 percent of the weight of the asphalt.

Vacuum distillation selectively removes higher volatility, lower molecular weight hydrocarbons, thereby resulting in a concentration of higher molecular weight (lower volatility) components in asphalt. During air-blowing, cyclics are converted to resins, which are in turn converted to asphaltenes. As the asphaltene content increases, the stiffness of the asphalt increases. Oxidized asphalts typically have higher asphaltene contents than non-oxidized asphalts.

Under ambient conditions, the physico-chemical properties of asphalt change very slowly over time. This may be due to the formation of additional asphaltenes by oxidation.

Studies of the rate of oxidation reactions have shown that the hetero-atoms, sulfur, oxygen, and nitrogen, occur largely in stable configurations, probably rings. X-ray diffraction points to the presence of naphthenic groups as well as aromatic and paraffinic groups. Other studies have led to the conclusion that oxygen and nitrogen, as well as vanadium and/or nickel, are combined in some form in the asphaltene structure.

Polarity

Polarity is the asymmetric distribution of charge within a molecule. During the self-assembly (organization) process, a preferred structure is created that is held together by electrostatic and other non-covalent forces. These forces are much weaker than covalent bonds, which means that they will break before the covalent bonds under conditions of increased stress and temperature. These electrostatic bonds break and reform relatively easily, unlike covalent bonds which, when they break, form new molecules.

At in-service temperatures, covalent bonds remain intact (except for those experiencing oxidation. So, when the non-covalent bonds break, the structure reorganizes without creating new molecules. However, that does not mean that the reorganized structure will have the same properties as the structure before reorganization occurs. As the structure becomes more stable, the material becomes more resistant to deformation.

Oxidation

As the asphalt molecules react with oxygen, new polar sites are formed which increases the tendency to self-assemble and thereby increases the stiffness of the asphalt. This type of oxidation can occur at the refinery, during mixing at an asphalt mixture production facility, and/or in service in the pavement. Because asphalts come from different crude sources, they have different molecular distributions and different attractive strength connecting the molecules. Therefore, the rate of stiffening can be different for different asphalts. Figure 2.1 shows the molecular distribution for three asphalts used in the Strategic Highway Research Program (SHRP). The graph (TRB Circular 499) was generated using size exclusion chromatography (SEC) which separates materials by molecular (or multimolecular) size.

Note that the three asphalts in Figure 2.1 have very different molecular profiles despite a similarity in their physical properties. Asphalt AAM has a high amount of large molecules, but it also has continuous molecular size. Asphalt AAG has a low amount of large molecules, but it has a high amount of small molecules. Asphalt AAK is the reverse, having a high amount of larger molecules but a low amount of small molecules. Neither AAG nor AAK has a continuous molecular size, meaning they each lack a significant amount of intermediate-sized molecules.

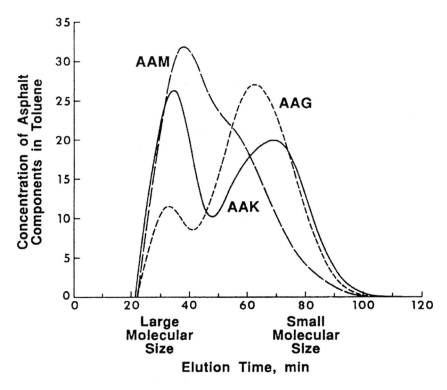

Size exclusion chromatogram of three asphalts

Figure 2.1 SEC of asphalts

How do the molecular sizes affect the properties of the asphalt after oxidation? With a high amount of smaller molecules, AAG has the ability to withstand the development of more polar molecules (as happens during oxidation) without changing its physical properties as significantly as another asphalt with more continuous molecular sizes (AAM). On the opposite hand, AAK—which has a high amount of larger molecules—is already dominated by the polar species. Thus oxidation will cause the polar molecules to be greater and more dominant, leading to a greater increase in stiffness. This behavior is illustrated in Figure 2.2 which shows the effect of aging on stiffness for the three asphalts. The RTFO and PAV aging (discussed later) serve to progressively age the asphalt from its original condition. Thus, the further to the right on the x-axis, the more highly aged (oxidized) the asphalt binder.

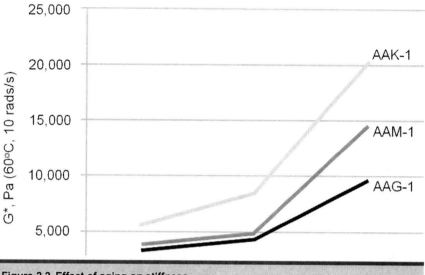

Figure 2.2 Effect of aging on stiffness

PAHs

Crude oil contains trace amounts of polynuclear aromatic hydrocarbons (PAHs), which end up in asphalt at ppm levels. The maximum temperatures involved in the production of asphalt are typically between 350 and 500°C (660 and 930°F). These temperatures are not high enough to initiate significant PAH formation, which requires pyrolysis or combustion and typically occurs at temperatures above 500°C (930°F). Vacuum distillation, the principal refinery process used for the manufacture of asphalt, removes the majority of compounds with lower molecular weight with lower boiling points—including PAHs with 4–6 unsubstituted, fused rings. For more information about PAHs please see *IS-230 The Bitumen Industry – A Global Perspective*, available from the Asphalt Institute.

CHAPTER 3

Asphalt physical properties and characteristics

Physical properties
Consistency
Purity
Safety

How asphalt behaves
Fundamental vs. empirical properties

Sampling, heating and splitting asphalt binders
Sampling
Heating and splitting asphalt binder samples
Changes in behavior with storage and handling

Physical properties

For engineering and construction purposes, three properties or characteristics of asphalt are important: consistency (also often called viscosity), purity, and safety.

Consistency

Asphalts are thermoplastic materials, meaning that they liquefy when heated and solidify when cooled. They are characterized by their consistency or ability to flow at different temperatures. Consistency is the term used to describe the viscosity or degree of fluidity of asphalt at any particular temperature. The consistency of an asphalt binder varies with temperature; therefore, it is necessary to either define an equivalent temperature or an equivalent consistency when comparing the temperature-consistency characteristics of one asphalt binder with another. In some specifications, asphalt binders are graded based on their consistency at a standard temperature; in other specifications, asphalt binders are graded based on the temperature at which they meet a standard consistency.

Asphalt hardens when exposed to air in thin films at elevated temperature. This occurs during the mixing process with aggregate in a hot mix asphalt mixing facility. When asphalt hardens, this means that the consistency or viscosity of the asphalt increases. While this increase in stiffness is expected and allowable to a certain extent, a great deal of damage through hardening can be caused to the asphalt binder by carelessly overheating the asphalt during mixing. Excessive hardening may cause the asphalt pavement to experience early cracking failures, significantly shortening the life of the finished asphalt pavement. Common methods for determining the consistency of an asphalt binder and its importance in standard specifications will be discussed in following paragraphs.

Purity

Asphalt binders are composed almost entirely of bitumen, which by definition is entirely soluble in carbon disulfide (or trichloroethylene). Refined petroleum asphalts are almost pure bitumen and are usually more than 99.5 percent soluble in carbon disulfide. Organic material and impurities, if present, are chemically inert.

Safety

Asphalt foaming is a safety hazard, and specifications usually require that asphalt not foam at temperatures up to 347°F (175°C). Normally, asphalt is free of water or moisture as it leaves the refinery. However, transports loading asphalt may have some moisture present in their tanks. If any water is inadvertently present, that water will become steam when the hot asphalt is added to the tank; the steam will cause the asphalt to foam, greatly increasing the volume of the asphalt in a short period of time. This rapid volume increase can easily result in a container (i.e., truck, sample container) overflowing and creating a burn hazard for anyone working near the overflowing container.

Asphalt binder, if heated to a high enough temperature, releases fumes that flash in the presence of a spark or open flame. The temperature at which this occurs is called the flash point and is well above the temperatures normally used in paving operations. However, to be certain of an adequate margin of safety, the flash point of the asphalt is measured and controlled.

How asphalt behaves

Asphalt binders are viscoelastic materials, meaning that they exhibit both viscous (fluid) behavior at high temperatures and elastic (solid) behavior at low temperatures. This behavior depends on both temperature and rate of loading. As shown in Figure 3.1, the amount an asphalt binder will flow could be the same for 1 hour at 60°C or 10 hours at 25°C. In other words, the effects of time and temperature are related in the measurement of flow properties of an asphalt binder. Thus, the flow of an asphalt binder at high temperature loaded for a short period may be equivalent to the flow occurring at a lower temperature for a longer loading duration. This useful concept is known as time-temperature superposition and it applies to linear viscoelastic materials.

At high temperatures, asphalt binders typically act as Newtonian fluids or non-Newtonian behavior as shear rate is changed. Newtonian fluids have a linear relationship between resisting force (shear stress) and relative velocity (rate of shear strain). In other words, if you apply twice the force to a Newtonian fluid, it will move twice as fast. Air, water, and hot asphalt (at temperatures greater than 60°C) are common Newtonian fluids. For these materials, viscosity (μ) is constant regardless of shear rate as shown in Figure 3.2.

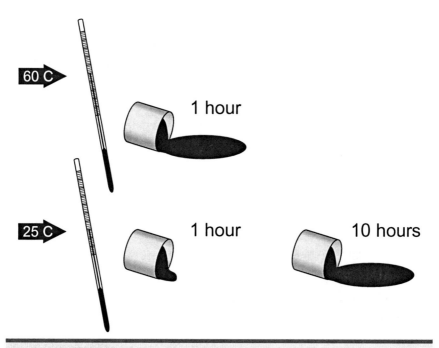

Figure 3.1 Time-temperature superposition

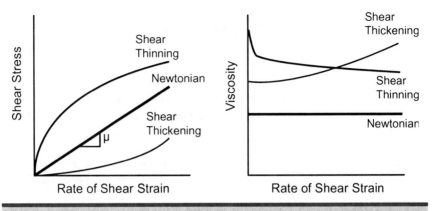

Figure 3.2 Behavior of asphalt binder as shear rate changes

Some asphalt binders, particularly modified asphalts, may exhibit non-Newtonian behavior at higher temperatures. Some asphalt materials may exhibit pseudoplastic or "shear thinning" behavior. This type of behavior, demonstrated in paints, is characterized by a decrease in viscosity as shear rate is increased (Figure 3.2). In other words, the faster you stir the material, the thinner (more fluid) it gets. At moderate temperatures, asphalt exhibits "shear thinning" behavior since its viscosity decreases at increasing shear rates.

Another type of non-Newtonian behavior, although less common in asphalt materials, is dilatant or "shear thickening" behavior. This type of behavior, demonstrated in clay slurries, is characterized by an increase in viscosity as shear rate is increased (Figure 3.2). In other words, the faster you stir the material, the stiffer (less fluid) it gets.

In cold conditions, such as winter temperatures, or when subjected to rapidly applied loads, such as fast-moving trucks, an asphalt binder behaves like an elastic solid. Elastic solids are like rubber bands; when loaded they deform, and when unloaded, they return to their original shape. If stressed beyond the material capacity or strength, elastic solids may break. Asphalt binders are elastic solids at low temperatures, and they may become too brittle with age and crack when loaded. For this reason, low-temperature cracking sometimes occurs in asphalt pavements during cold weather. In these cases, the load is applied by internal stresses that accumulate in the asphalt pavement as a result of low temperatures instead of vehicles. As the temperature decreases, the asphalt binder cementing the aggregates contracts, leading to a buildup of stresses. If the internal stresses in the asphalt binder exceed the binder's strength, the interface between the aggregates will break, resulting in the formation of a crack. In other words, if the rate of contraction exceeds the rate of flow, a crack will occur.

At intermediate temperatures, asphalt binders exhibit the characteristics of both viscous liquids and elastic solids. When heated, asphalt acts as a lubricant, allowing the aggregate to be mixed, coated, and tightly compacted to form a smooth, dense surface. After cooling, the asphalt acts as the glue to hold the aggregate together in a solid matrix. In this finished state, the behavior of the asphalt is viscoelastic, meaning it has both elastic and viscous characteristics, depending on the temperature and rate of loading.

At a constant rate of loading, asphalt behaves like an elastic solid at very low temperatures and like a viscous fluid at very high temperatures. This is illustrated in Figure 3.3. At very low temperature the stiffness reaches a point where it doesn't change very much for a given decrease in temperature. This asymptote is referred to as the glassy modulus. For asphalt binders, the glassy modulus is typically 1×10^9 Pa (or 1 gigaPascal). By contrast, at high temperature the asphalt binder stiffness reaches another asymptote referred to as the viscous asymptote. At this point on the curve, any change in temperature (or loading frequency) results in the same amount of change in stiffness.

Because asphalt binders are composed of organic molecules, they react with oxygen from the environment. This reaction is called oxidation, and it changes the structure and composition of asphalt molecules. Oxidation causes the asphalt binder to become even stiffer and more brittle, leading to an effect called oxidative hardening. Oxidative hardening happens at a relatively slow rate in a pavement, although it occurs faster in warmer climates and during warmer seasons. Because of this hardening, old asphalt pavements may be more susceptible to cracking than newer pavements. However, it should be noted that improperly compacted asphalt pavements might exhibit premature oxidative hardening. In this case, inadequate

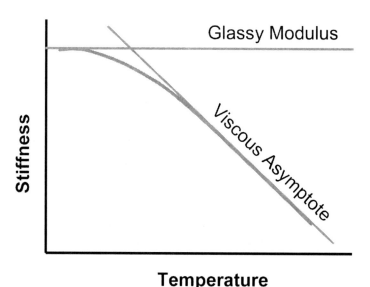

Figure 3.3 Asphalt binder stiffness as a function of temperature

compaction leaves a higher percentage of interconnected air voids, which allows more air to penetrate the asphalt mixture, leading to more oxidative hardening early in the life of the pavement.

In practice, a considerable amount of oxidative hardening occurs before an asphalt mixture is placed. At the hot-mixing facility, asphalt binder is added to the hot aggregate and the mixture is maintained at elevated temperatures for a time. Because the asphalt cement exists in thin films covering the aggregate at high temperature, the oxidation reaction may occur at a relatively fast rate.

The rate at which oxygen reacts with the asphalt binder depends on temperature and the availability of the oxygen to the individual asphalt molecules. As a rough rule of thumb, the rate at which oxidation occurs doubles for every 10°C increase in temperature. This increase is compounded, meaning that the rate quadruples for a 20°C increase in temperature. This behavior is illustrated in Figure 3.4.

In the example above and in Figure 3.4, the theoretical doubling is based on the assumption that oxygen is readily available to react with the asphalt molecules. If oxygen is blocked, as in a tightly sealed container, then the oxidation is negated. An oxidation rate that doubles as shown in Figure 3.4 is rarely seen in practice since, for this to occur, the asphalt must be in very thin films. This illustrates two key points: (1) oxidation rate is highly dependent on a combination of temperature and time; and (2) oxidation rate is reduced if the access to oxygen is limited.

As asphalt binders are heated, lighter weight molecules start to evaporate. This loss of weight during heating is called volatilization. Like oxidation, volatilization increases with temperature. At room temperature and in-service temperatures, asphalt binders experience very little volatilization. When heated to temperatures used in the laboratory for pouring samples

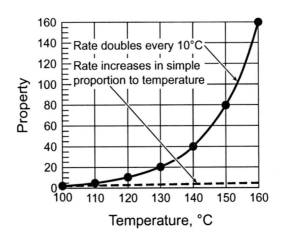

Figure 3.4 Effect of temperature on property change

and in the mixing plant for making asphalt concrete mixtures, measurable volatilization can occur. The loss of the lighter weight molecules during volatilization means that the asphalt binder has heavier molecules remaining, thus leading to an increase in stiffness.

At room temperature, a different kind of reaction can occur that results in an increase in asphalt binder stiffness. This reaction is called steric hardening. In practice, steric hardening is the slow molecular association, or structuring, that naturally occurs between the various polar and nonpolar molecules in an asphalt binder. It occurs as the asphalt binder is cooled to room temperature and continues to develop over a long period of time.

Steric hardening is very asphalt-specific since it is strongly dependent on the chemical composition of the asphalt binder. It may be observed as a significant effect for one asphalt binder, while not being observed at all for a different asphalt binder.

The amount of steric hardening that occurs within one day may be sufficient to cause test results to be significantly different depending on how long the sample had been at room temperature before testing. This is one reason why many test methods require that testing be conducted within a specified time period after the sample has been poured.

Fortunately, unlike oxidation, steric hardening is a reversible effect. To reverse the effects, the asphalt binder simply needs to be heated and stirred until sufficiently fluid to pour. This process, called annealing, destroys the molecular associations generated during steric hardening.

At low temperature, yet another kind of reaction can occur that results in an increase in asphalt binder stiffness. This reaction is called physical hardening. Unlike steric hardening, which is associated with molecular association, physical hardening appears to be related to densification of the asphalt binder and crystallization of wax present in the sample that occurs as the temperature is decreased below room temperature. Because wax can appear in many different forms, not all wax compounds will contribute equally to the physical hardening mechanism. Asphalt binders that exhibit significant steric hardening may exhibit little physical hardening.

The physical hardening process begins as soon as the asphalt binder is cooled below room temperature and continues for extended periods of time. The effect of physical hardening, an increase in stiffness, increases as the temperature is lowered. The highest temperature at which physical hardening occurs is asphalt-specific, but generally harder asphalt binders show signs of physical hardening at warmer temperatures than softer asphalt binders.

Figure 3.5 illustrates physical hardening in three different asphalt binders maintained at low temperature for as long as 100 hours. As shown in the figure, the rate and amount of physical hardening are asphalt binder specific. It should also be noted that the hardening that occurs in 24 hours, a two-fold increase in stiffness, may be comparable to the increase in stiffness due to oxidation from in-service aging.

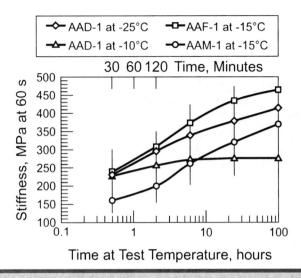

Figure 3.5 Effect of physical hardening

Like steric hardening, physical hardening is a reversible effect; it can be reversed by heating the asphalt binder to room temperature. Also like steric hardening, physical hardening can be controlled by tightly controlling the amount of time that the test specimen is allowed to remain at low temperature before testing.

Fundamental vs. empirical properties

Fundamental mechanical properties used in asphalt binder specifications include the coefficient of viscosity, complex modulus, and stiffness. What makes these properties fundamental? Fundamental properties are based on units of stress and strain and are independent of sample size and dimensions of test equipment.

Empirical mechanical properties that have been used in the past in asphalt binder specifications include penetration, ductility, and softening point. What makes these properties empirical? Empirical properties are not based on stress and strain and are dependent on sample size and test dimensions and conditions. If the penetration needle changes size or shape, then the measured value for penetration will change. If a ductility test is conducted with a different size mold, the reported ductility value will change.

In the characterization of an asphalt binder's mechanical properties, technologists expect that the mechanical property measured in the lab will somehow relate to the performance of the asphalt binder in practice. The relationship with performance can only be accomplished through a series of complex equations which depend upon properties that describe the mechanical behavior of the asphalt binder. As such, it is important that these properties are fundamental mechanical properties and not empirical properties.

Sampling, heating and splitting asphalt binders

If the sampling and handling of an asphalt binder is poor, then it is very likely that the test results will also be poor. Care is required in the field and laboratory to ensure that the sample of asphalt binder is representative and that the properties are unaffected by handling before testing.

Sampling

The primary reference for proper field sampling is AASHTO T 40, *Standard Method of Practice for Sampling Bituminous Materials*. While many asphalt binder testing technicians may never sample asphalt binders in the field, it is nevertheless important that they have an understanding of how the sampling practice can affect test results.

The purpose of any field sampling program is to ensure that any sample taken is representative of the total material from which it is taken. This is termed a "representative sample." Rather than trying to show examples of representative samples, it is easier to illustrate examples of nonrepresentative samples, such as:
- an asphalt binder sample poured into a silicone rubber mold one evening and tested the next day;
- a sample always taken at the point of delivery when the transfer of material has just started; and
- a sample from a mixing plant that is always taken from the tank at startup.

In quality control operations, reference is often made to "random sampling" or a "random sample." A random sample may be defined as a sample that has "a chance of being collected that is equal to the chance of any other sample that could be collected." For aggregates, a sample that is always collected at the bottom of a stockpile is not necessarily representative of the entire stockpile. When tested, these samples may give a misleading characterization of the stockpile properties. Samples taken at random locations throughout the stockpile have a much better chance of properly representing the properties of the whole stockpile.

To be a representative sample, it must be obtained in an appropriate manner. This procedure is described in AASHTO T 40. Individual quality control plans may also contain additional provisions to ensure proper sampling. Some important aspects of proper sampling include:
- Sampling personnel: Sampling should be performed under the supervision of a qualified person to prevent the violation of proper sampling procedure, leading to the procurement of a poor sample.
- Sample location: Asphalt binder samples should be obtained in-line, with proper care taken to flush the lines before the sampling

occurs. At a minimum, one gallon of asphalt binder should be flushed from the lines (and wasted) before taking the test sample. The practice of obtaining a sample from a tank by "dipping" should be avoided.

- Sample containers: Asphalt binder samples should be collected in new metal containers with friction lids (e.g., paint cans). The cans should be unlined and free of oil or other contaminants (in some cases an oily residue is left in the cans when they are manufactured). Friction lids should be placed loosely on the cans until the sample has cooled to the touch. At that point the lids can be firmly tapped in place.

    Screw-top cans should not be used because it is typically difficult to remove the covers and properly stir and pour the asphalt binder. If used, screw tops should be screwed tightly to seal the can after the asphalt binder has cooled to room temperature.

    It is important that asphalt binder samples are obtained using a container of the appropriate size for the testing need. For normal asphalt binder physical property testing, a quart (liter) container will provide sufficient sample to conduct all the physical testing needed for a typical specification. For additional physical or chemical testing, more quart samples may be needed.

    Quart-sized containers are preferred for sampling and testing because the sample size is smaller meaning that hot samples aren't as heavy (they are easier to handle and safer); they heat relatively quickly to pouring temperature; and they are easy to pour from. In general, one-gallon (4-liter) and five-gallon (20-liter) containers are not preferred for sampling. An exception is when the samples will be used for mixture design testing. In this instance, a large-sample size is needed and larger containers are more efficient. If large containers are used, the technician should heat the entire sample, stir thoroughly, and immediately separate the entire sample into multiple one-quart containers so that reheating can be minimized.

- Proper filling: When an asphalt binder is poured into a sample container, it should always be poured so that the container is as full as practical. Some free space is needed at the top of the container so that the asphalt binder can be stirred and poured without spilling. While messy, the more important negative aspect about spilling is that it tends to obliterate labels and can be a safety hazard. Rules of thumb for proper filling height are shown in Figure 3.6.

- Proper labeling: As with all samples, proper labeling is very important. This is particularly true for asphalt binder samples where the asphalt binder could drip down the side of the sample container and partially or completely obscure sample information. As noted above, caution is needed when emptying or filling a container so that the asphalt binder does not obscure the label and/or sample information. Placing a sample label over the seam of the sample container usually helps to minimize label damage since technicians often avoid pouring over the seam.

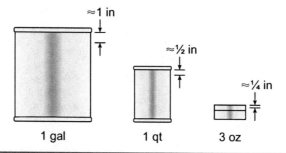

Figure 3.6 Sample filling

Sample information and labels should always be placed on the side of the sample container. It is far too easy to be in the process of testing several samples and realize that the lids—with all the sample information—have been misplaced or jumbled.

Heating and splitting asphalt binder samples

Asphalt binder samples must be heated in the laboratory to reduce the size of the samples and to prepare test specimens. Heating an asphalt binder will inevitably change its properties. To minimize property changes in the asphalt binder, several precautions can be taken:
- Heat the asphalt binder container in an oven. Avoid the use of hot plates and other heating devices that allow the heating elements to be in direct contact with the sample container.
- Heat the asphalt binder sample at the lowest temperature for the shortest period of time until it is sufficiently fluid to pour. Sample preparation guidelines are provided in many of the test procedures and may also be provided by the manufacturer.
- Avoid heating the asphalt binder in thin films, as will inevitably occur with shallow or nearly-empty containers.
- Loosely cover the sample container during heating. It is important that the container cover is loose enough so that pressure is relieved in the container as it is being heated. Although some will argue that the presence of a cover does not affect oxidation or volatilization, a sample that is loosely covered will, at the least, be protected from contamination.

Although the required amount of heating will vary depending on the stiffness of the asphalt binder, the common sense approach of heating as little as possible for the shortest time possible is the best guideline.

Heating a sample using a direct flame—such as heating a container over a Bunsen burner—is discouraged because of the possibility of localized overheating. However, one practice used by some technicians in recent years, particularly for oxidized asphalt binder samples, has been to pass a flame lightly over the sample surface to remove air bubbles. This practice will probably do little damage to the physical properties of the asphalt binder as long as the flame is passed very rapidly over the surface in just a few passes.

In many labs, hot plates have been used to heat samples. Although an oven is preferred, a hot plate can be used if the sample container is not placed directly on the surface. The best way to accomplish this is through the use of a sand bath. A sand bath (Figure 3.7) allows any localized hot spots to be mitigated and can serve to help maintain the sample at a proper temperature for specimen preparation.

In all cases, regardless of the heating medium, the heating of an asphalt binder sample should be accompanied with stirring to ensure sample homogeneity. This is particularly true for many modified-asphalt binders, although simple heating and stirring may not be sufficient to return the sample to its as-produced condition.

Unless the asphalt binder sample is the proper size for testing, the sample will have to be split to the appropriate size. Reheating samples more than once should be avoided. When a sample arrives at the testing laboratory, best practice dictates that the sample be heated once and split into smaller containers that can be used once for future testing and discarded (Figure 3.8).

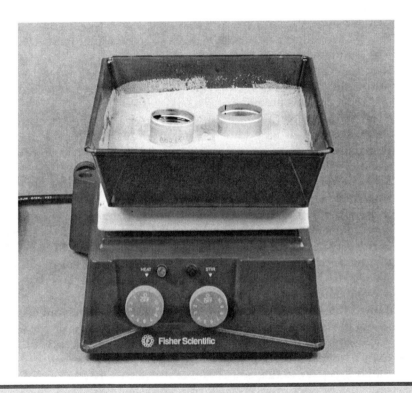

Figure 3.7 Heating a sample on a hot plate

Figure 3.8 Splitting a sample

Changes in behavior with storage and handling

The physical properties of asphalt binders can be affected in a number of ways between the time they are sampled and the time testing is completed. In the laboratory, the physical properties can be affected by the following:
- Prolonged exposure to air at room temperature: Damage may occur as a result of prolonged exposure of the asphalt binder sample to air as a result of damaged or partially open cans.
- Heating during sample splitting: As discussed previously, the act of dividing the contents of a larger container into one or more smaller containers is referred to as "splitting." As the sample is heated, oxidation and volatilization can occur. This is best mitigated by keeping the temperature as low as practical and the time as short as practical when heating the sample.
- Steric hardening: As discussed previously, steric hardening can be reversed.
- Physical hardening: As discussed previously, physical hardening can be reversed.
- Polymer degradation: Degradation can occur with some polymers if the asphalt binder is heated excessively.
- Polymer separation: Phase separation between the asphalt and polymer system can occur during storage at higher temperatures.

Supplier recommendations for handling should be followed whenever possible. Some modified-asphalt binders have specific handling instructions as they are heated for testing. Recommendations may include minimum or maximum heating temperatures, a stirring protocol to ensure homogeneity, and/or annealing instructions.

Asphalt binders that are stored at room temperature in airtight, mostly-filled containers with little free space will experience little change in physical properties over time. Storage at low temperature is not necessary. However, it is important to note that steric hardening is likely to have occurred over time and the sample will need to be annealed before it can be tested.

CHAPTER 4

Asphalt cements

In the past, the term "asphalt cement" was used by the asphalt industry in the United States to represent straight-run, or unmodified, paving grade asphalts. With the increasing use of modification, the term "asphalt binder" was later adopted to be more inclusive of all asphalts—including those that had been chemically and/or physically modified.

To verify that an asphalt cement had the desired physical characteristics (consistency, purity, and safety), it was necessary to develop tests and specifications that related to those characteristics. Some of the tests described have existed since petroleum asphalt began to be widely used in the early 1900s. In this chapter, tests are separated as historical tests (those tests commonly used until the 1990s) and PG System tests (those tests comprising the Performance-Graded Asphalt Binder Specification produced from the research of the Strategic Highway Research Program).

(Author's note: tests are described in the following sections so that the reader can understand the general procedures. The synopsis of a given test is in no way intended to replace the actual test procedure. Please refer to appropriate standards in AASHTO and ASTM.)

Historical tests
  Viscosity
  Penetration
  Flash point
  Aging
  Ductility
  Solubility
  Specific gravity

Specifications for asphalt cements
  Penetration-graded specification
  Viscosity-graded specification

■ PG (Performance Graded) asphalt binder tests
Rotational viscometer
Dynamic Shear Rheometer (DSR)
Bending Beam Rheometer (BBR)
Direct Tension Tester (DTT)
Aging procedures

■ PG asphalt binder specification
Advantages compared with older specifications
Assumptions in the specification
How the PG specification addresses pavement performance
Grading an asphalt binder in the PG specification system
Performance-Graded asphalt binder using the Multiple Stress
  Creep Recovery (MSCR) test

Historical tests

Viscosity

Viscosity is a fundamental property that describes the resistance of a liquid to deformation or flow based on the internal friction of its molecules. Fluids that have a high viscosity at a given temperature would be expected to have a high resistance to deformation. Viscosity is expressed as the ratio of shear stress to the rate of shear in the fluid. In SI units, viscosity is reported in Pascal-seconds (Pa-s). In English units, viscosity is reported in poise (P) and is equal to 0.1 Pa-s.

In the testing of asphalt cements, three different viscosity tests may be performed: absolute, kinematic, and apparent. The absolute and apparent viscosity tests are conducted at 140°F (60°C). The kinematic viscosity test is conducted at 275°F (135°C). Before first use, each viscometer is calibrated by using standard oils with known properties. Calibration constants are then developed for each viscometer and used as described later in this chapter. Viscometers are usually calibrated by the manufacturer, and calibration constants are furnished with each tube.

Absolute viscosity

In the absolute viscosity test, two viscometer tube types are most commonly used: the Asphalt Institute vacuum viscometer (shown in Figure 4.1) and the Cannon-Manning vacuum viscometer. The absolute viscosity test equipment consists of a thermostatically-controlled, constant temperature bath (Figure 4.2) in which viscometer tubes may be placed. Water is often used as the liquid medium in the bath because the test temperature is below its boiling point.

To conduct the absolute viscosity test procedure, preheated asphalt is poured into the large side of the viscometer until its level reaches the filling line (Figure 4.3). The filled viscometer tube is then placed back in an oven operating at the same temperature used to heat the asphalt binder for a short time. Finally, the viscometer is placed in the constant temperature bath for a prescribed time to allow the asphalt to reach an equilibrium temperature of 140°F (60°C).

At 140°F (60°C), a paving-grade asphalt binder has too high a viscosity to flow readily through a capillary tube viscometer. Therefore, the absolute viscosity test requires the use of a vacuum pump to apply a consistent partial vacuum to the small side of the viscometer to induce flow.

After the bath, viscometer, and asphalt binder have stabilized at 140°F (60°C), the prescribed vacuum is applied and the time in seconds required for the asphalt binder to flow between two timing marks (delineating a section of capillary tube also called a "bulb") is precisely measured. Multiplying the measured time to flow through this bulb by the calibration

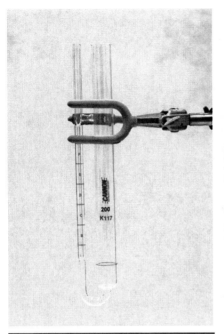

Figure 4.1 AIVV Tube

Figure 4.2 Absolute viscosity bath with tubes

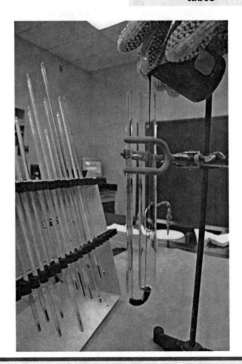

Figure 4.3 Pouring absolute viscosity tubes

constant for the particular bulb of the viscometer gives the value for viscosity in *poises,* the standard unit for measurement of viscosity.

There are several different sizes of Asphalt Institute vacuum viscometers, with different-size capillary tubes. Common sizes used for paving-grade asphalt binders are identified as 50, 100, and 200 tubes. Higher-number tubes are used for asphalt binders with greater viscosity. There are also 400 and 800 tubes, usually followed with an R designation, that are used for roofing asphalts (hence the R designation) and very high viscosity paving-grade asphalts. Each viscometer also has several timing marks, delineating several bulbs. To ensure the repeatability of viscosity measurements, the test procedure requires that the time to flow through a bulb (between successive timing marks) should be at least 60 seconds before calculating viscosity (multiplying time by the calibration constant for the appropriate bulb). If the time to flow through the last bulb in a viscometer is less than 60 seconds, then the asphalt should be retested using the next smaller-numbered tube (i.e., a 100 instead of a 200 tube).

Kinematic viscosity

Unlike viscosity measurements at 140°F (60°C), paving-grade asphalt cements are sufficiently fluid at 275°F (135°C) to flow through capillary tubes under gravitational force alone. This type of viscosity is called kinematic viscosity. To determine viscosity at this high temperature, a vacuum is not required, but a different type of viscometer is used. The most commonly used is the Zeitfuchs cross-arm viscometer (Figure 4.4).

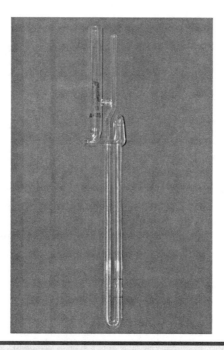

Figure 4.4 Cross-arm viscosity tube

Figure 4.5 Kinematic viscosity bath with tubes

As with absolute viscosity, each kinematic viscometer tube is also calibrated using standard calibrating oils. Calibration constants are developed for each viscometer and used as previously described. As with absolute viscosity, viscometers are calibrated by the manufacturer and calibration constants are furnished with each tube.

Because kinematic viscosity tests are conducted above the boiling point of water at 275°F (135°C), a suitably clear oil must be used as the bath medium. The viscometer is mounted in the bath (Figure 4.5), and asphalt is poured into the large opening until it reaches the filling line. The system is then allowed to reach equilibrium temperature as before. A slight pressure is applied to the large opening, or a slight vacuum is applied to the small opening, to start the asphalt flowing over the siphon section just above the filling line. The asphalt will then flow downward in the vertical section of capillary tubing by gravitational force. A timer is started when the asphalt reaches the first timing mark and stopped when it reaches the second. The timer interval, multiplied by a calibration factor for the viscometer, gives the kinematic viscosity in centistokes.

There are several different sizes of the crossarm viscometers, with different size capillary tubes. Common sizes used for paving-grade asphalt cements are identified as 5, 6, and 7 tubes. Higher-number tubes are used for asphalt binders with greater viscosity. As with the absolute viscosity test, in order to ensure the repeatability of viscosity measurements, the time to flow between the timing marks should be at least 60 seconds. If the time to flow is less than 60 seconds, then the asphalt should be retested using the next smaller-number tube.

As an alternative, the rotational viscosity test may also be used to determine the viscosity of an asphalt binder at 275°F (135°C). The rotational viscosity equipment and test procedure are described later in this chapter.

Note that the kinematic viscosity at 275°F (135°C) is expressed in centistokes while the absolute viscosity at 140°F (60°C) is expressed in poises. Gravity induces the flow in the kinematic viscosity test (results in centistokes), and the density of the asphalt affects its rate of flow through the capillary tube. In the absolute viscosity test (results in poises), flow through the capillary tube is induced by a partial vacuum where gravitational effects are negligible. The units of the two tests, poises and stokes or centipoises and centistokes, are related to each other by the density of the asphalt.

Viscosity measurements made at the two test temperatures are sometimes plotted on a viscosity-temperature graph with a line connecting the two points. The slope of this line is an indication of the temperature susceptibility of the asphalt—the greater the slope of the line, the greater the temperature susceptibility. Asphalts that have high temperature susceptibility may be softer at hot temperatures and harder at cold temperatures—a bad combination for successful performance at in-service pavement temperatures.

Additionally, the temperature-viscosity relationship of an asphalt binder can be important in determining appropriate temperature ranges for laboratory mixing and compaction of asphalt mixtures. This topic is discussed in Chapter 10.

Apparent viscosity

In determining the absolute viscosity of an asphalt binder by ASTM D 2171 (AASHTO T 202), the assumption is made that the viscosity remains constant regardless of the shear rate. If the shear stress is proportional to the shear rate, then the viscosity is constant and the fluid is considered to be Newtonian (Chapter 3, Figure 3.2). Unfortunately some modified asphalt binders can exhibit significant non-Newtonian behavior meaning that the measured viscosity is a function of the shear rate. The apparent viscosity test procedure allows the user to better understand the behavior of non-Newtonian asphalt binders.

To conduct the apparent viscosity test, a vacuum capillary viscometer is used and the preheated asphalt sample is poured into the viscometer tube until its level reaches the filling line. The filled viscometer tube is then placed back in an oven operating at the same temperature used to heat the asphalt binder for a short time. Finally, the viscometer is placed in the constant temperature bath for 30 minutes to allow the asphalt to reach an equilibrium temperature of 140°F (60°C). After 30 minutes, a vacuum is applied to the viscometer tube to start the flow of asphalt. The time for the asphalt binder to flow between successive timing marks is recorded. Unlike the absolute viscosity test procedure which may end once the time between two timing marks exceeds 60 seconds, the apparent viscosity procedure collects data from all bulbs within the tube. The apparent viscosity for each bulb can be calculated using the viscometer calibration constant and the measured time. The shear rate associated with each apparent viscosity calculation can also be calculated using the shear constant for the tube divided by the time required by the asphalt binder to flow through the bulb.

Once the test is complete and the apparent viscosity and corresponding shear rate are calculated, a log-log graph may be plotted (apparent viscosity versus shear rate) as shown in Figure 4.6. Unless otherwise noted, ASTM D 4957 recommends reporting the apparent viscosity at a shear rate of 1 s^{-1} (calculated from the graph). Ideally, the test will have produced apparent viscosity values that have corresponding shear rates greater than and less than 1 s^{-1}. If not, the curve may be extrapolated (not preferred, but acceptable for a short extrapolation) or the test can be rerun using a different viscometer tube to create different shear rates. Finally, the test procedure also provides for the option of changing the vacuum to create different shear rates. For asphalt binders with a lower apparent viscosity, the vacuum may be decreased. By contrast, for asphalt binders having a higher apparent viscosity, the vacuum may be increased. In either case, new calibration constants for each bulb will need to be calculated.

In Figure 4.6, the test data indicate that the asphalt binder has an apparent viscosity of 68,364 poises at a shear rate of 0.34 s^{-1} and an apparent viscosity of 29,704 poises at a shear rate of 1.35 s^{-1}. Interpolation between the two data points indicates that the apparent viscosity at 1 s^{-1} is 35,612 poises.

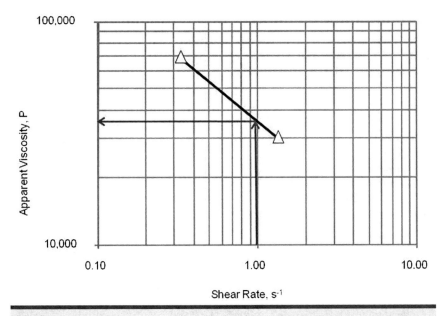

Figure 4.6 Apparent viscosity graph

Note that the behavior shown in the figure—viscosity decreasing as shear rate increases—is an example of shear-thinning behavior (see Chapter 3).

Penetration

The penetration test is one of the oldest, empirical tests for measuring the consistency of an asphalt binder. The test was originally devised by H.C. Bowen in 1888 and later improved by A.W. Dow. Figure 4.7 shows the standard penetration test equipment.

To conduct the penetration test, an asphalt binder sample is heated to an appropriate pouring temperature and poured into a test container—usually a 3-ounce tin (for standard paving-grade asphalt binders). After a specified conditioning period (tightly controlled to minimize the effects of steric hardening), the asphalt binder sample is brought to the standard test temperature of 77°F (25°C) in a temperature-controlled water bath. The sample container is then placed in the penetrometer equipment. A needle of prescribed dimensions is attached to the penetrometer and suspended directly over the asphalt binder sample. A 50-gram weight is attached to the needle's loading platform so that the total weight used for loading is 100 grams (50-gram weight plus the needle assembly weight of 50 grams). The penetrometer is lowered until the needle tip just contacts the surface of the asphalt binder. The load is then released, allowing the weighted needle to penetrate the asphalt binder for 5 seconds. The distance that the needle penetrates into the asphalt binder is reported as the penetration value (Figure 4.8). This distance is reported in units of 0.1 millimeters, or decimillimeters (dmm).

A typical penetration test consists of three measurements made on the asphalt binder sample that are averaged to provide a single test value. Between each test, the needle is replaced with a clean needle and the sample container is rotated so that a different portion of the sample is tested.

Occasionally, the penetration test is conducted at a different test temperature, such as 41°F (4°C). In these instances, total load, penetration time, or both, may be varied.

Flash point

As a petroleum product, an asphalt binder will release combustible fumes when heated to sufficiently high temperatures. The flash point provides an indication of the temperature at which a heated asphalt binder sample will instantaneously flash in the presence of an open flame. It is important to note that this temperature is usually well below the temperature at which the material will support combustion. The temperature at which the material supports combustion is called the fire point. It is rarely, if ever, used in asphalt binder specifications. Additionally, at least one user agency has used the flash point test to determine the smoke point of the

Figure 4.7 Penetrometer

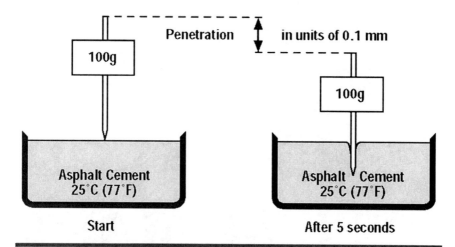

Figure 4.8 Penetration test

Figure 4.9 COC test

asphalt binder. As the name implies, this is the point at which the sample begins to exhibit visible smoke.

The most common test method used to determine the flash point of an asphalt binder is the Cleveland Open Cup (COC) flash point test (Figure 4.9). In this test, a brass cup is first filled with a specified volume of asphalt binder and is heated at a constant rate (Figure 4.10). A small flame is passed over the surface of the asphalt at timed intervals (Figure 4.11).

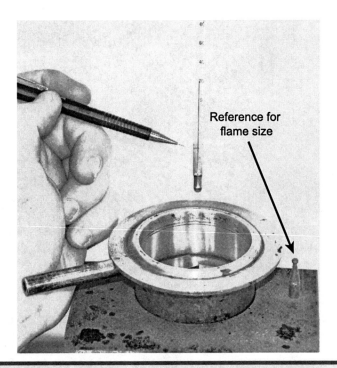

Figure 4.10 COC overview (manual)

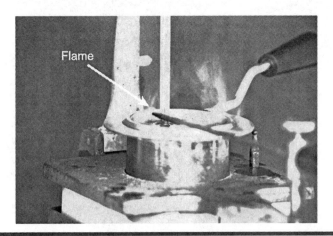

Figure 4.11 COC manual operation

When the flame passing over the sample surface causes an instantaneous flash (Figure 4.12), the temperature is recorded as the material's flash point.

The purpose of the flash point test is to measure the tendency of the asphalt binder to produce flammable vapors when heated under controlled laboratory conditions. As stated in the test method, flash point

Figure 4.12 COC flash

shall not be used to appraise the risk of a fire hazard, but instead it shall be one component in an overall strategy for fire risk management. The results of the test method are related to safety and not the quality of the asphalt binder.

Aging

Asphalt technologists have recognized that while the physical properties of some asphalt binders may be the same initially, the oxidation and volatilization that occurs during use in a hot-mix facility and subsequent placement on the road might lead to a different in-service performance. Stated more simply, not all asphalt binders age the same and this difference in aging could result in a difference in actual performance.

To address this issue, aging procedures were developed to subject an asphalt binder sample to hardening conditions that approximate the conditions that occur in normal, hot-mix facility operations. The Thin-Film Oven Test (TFOT) and Rolling Thin Film Oven Test (RTFOT) are two common short-term aging procedures. Historically, the TFOT had seen more widespread use than the RTFOT, which principally had been used in the western United States. This changed with the development of the Performance-Graded (PG) Asphalt Binder Specification, which now requires the use of the RTFOT. This will be discussed in more detail later.

To address the issue of volatilization—the process of lighter volatile fractions being removed from the asphalt binder due to heating at elevated temperatures—both the TFOT and RTFOT procedures incorporate a mass change determination. In this determination, the test container (pan or bottle, depending on the procedure) is weighed initially. It is weighed a second time after the asphalt binder sample is added and a third time after the completion of the test. The change in mass is calculated based on the weight of the sample before and after the aging procedure. It is expressed as a percentage of the mass before aging.

For most asphalt binders, the exposure to elevated temperatures and air results in volatilization and a loss of mass during the test. Usually, asphalt binders that are softer at room temperature might be expected to experience greater mass loss than stiffer asphalt binders. While uncommon, some asphalt binders experience a mass gain during the TFOT or RTFOT aging procedures. In these cases, the combination of heat and oxygen interacts in such a way with the asphalt binder that there is an increase in the mass.

Thin-Film Oven Test (TFOT)

The TFOT procedure is performed by pouring 50 grams of heated asphalt binder into a flat-bottom, circular sample pan with an inside diameter of 5.5 inches (140 millimeters) and a depth of 3/8 inch (10 millimeters). After pouring, the layer of asphalt binder is approximately 1/8 inch (3 millimeters) thick. As noted earlier, if a mass change determination is needed, then the sample pans are weighed before and after the addition of the asphalt binder. The number of sample pans needed depends on the amount of sample needed for further testing after aging.

To start the test, the sample pans are placed on a rotating shelf in a ventilated oven (Figure 4.13) operating at 325°F (163°C). The rotating shelf turns at a rate of approximately 5 to 6 revolutions per minute for a total testing time of 5 hours. After 5 hours, any sample pans being used for determining mass change are cooled to room temperature before the final weight is determined. The remaining pans are then poured and scraped into a single sample container for additional testing, such as viscosity or penetration.

Rolling Thin-Film Oven Test (RTFOT)

The RTFOT procedure has the same purpose as the TFOT, but equipment and test procedures are different. Many asphalt technologists believe the RTFOT procedure provides slightly more aging (i.e., stiffness or viscosity) than the TFOT procedure.

The RTFOT procedure is performed by pouring 35 grams of heated asphalt binder into a glass bottle (Figure 4.14). As noted earlier, if a mass change determination is needed, then two designated RTFOT bottles are weighed before and after the addition of the asphalt binder. These bottles are then used to determine mass change. The number of bottles needed depends on the amount of sample needed for further testing after aging.

Immediately after pouring, the bottle is turned to a horizontal position and rotated one full turn (Figure 4.15) to pre-coat the bottle before it is placed horizontally in a cooling rack (Figure 4.16) to cool for 60–180

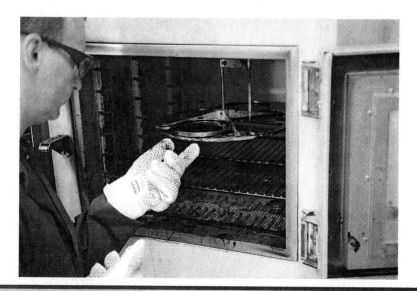

Figure 4.13 TFO

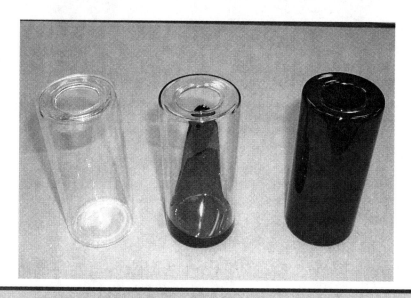

Figure 4.14 RTFO bottles

minutes. During this time, the cooling rack should be located away from heat sources (e.g., ovens) and in a draft-free location.

To start the test, the sample bottles are placed in a vertically rotating carriage in an oven (Figure 4.17) operating at 325°F (163°C). The bottle carriage rotates at 15 revolutions per minute for a total testing time of 85 minutes. During the rotation, a jet of air flowing at a rate of 4,000 milliliters

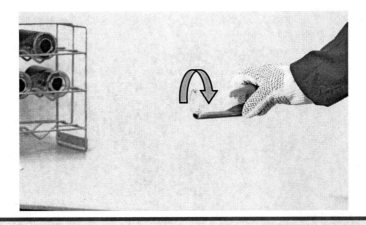

Figure 4.15 Bottle rotation

Figure 4.16 Cooling rack (ASTM)

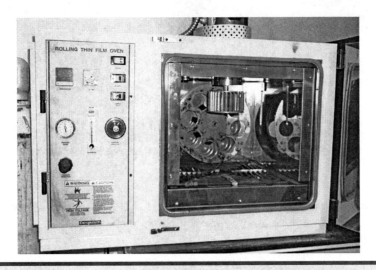

Figure 4.17 RTFO

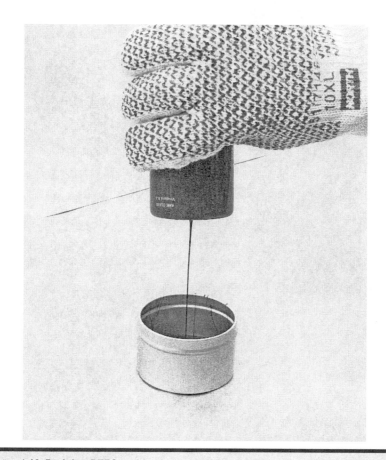

Figure 4.18 Draining RTFO residue into container

per minute blows into each bottle as it passes the bottom position in the carriage. After 85 minutes, any sample bottles being used for determining mass change are cooled to room temperature before the final weight is determined.

The RTFOT is usually conducted not only to determine mass change but also to produce aged residue suitable for further testing. As such, any remaining bottles are drained into a sample container (Figure 4.18) and scraped to remove the residue using a specialized tool (Figures 4.19 and 4.20). The aged residue can then be used for additional physical or chemical property testing.

Ductility

Ductility, as a physical property, has been considered an important characteristic of asphalt binders by some engineers. The presence or absence of ductility, however, is often considered more significant than the actual

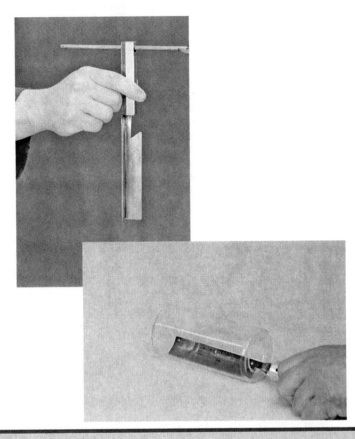

Figure 4.19 Scraping tool

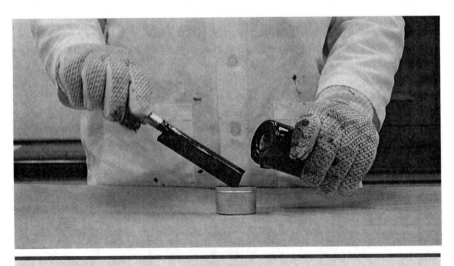

Figure 4.20 Scraping RTFO residue into container

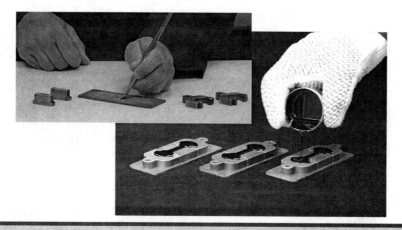

Figure 4.21 Mold preparation and pouring

Figure 4.22 Trimming ductility specimen

degree of ductility. This is because some asphalt binders having a high degree of ductility have also been found to be more temperature-susceptible.

The ductility test procedure is performed by first creating a standard mold that results in a cavity shaped like a dog bone. A heated sample of asphalt binder is then poured into the mold, slightly overfilling the mold (Figure 4.21). After a prescribed cooling period, the specimen is trimmed flush with the surface of the mold (Figure 4.22). The test specimen is then placed in the ductility water bath and conditioned to the desired test temperature, usually 77°F (25°C). The specimen is then loaded in the ductility machine and one end of the specimen is pulled away from the other at a specified rate of speed, normally 5 centimeters per minute, until the thread of asphalt connecting the two parts of the sample breaks

Figure 4.23 Ductility test

(Figure 4.23). The elongation at which the thread of material breaks is designated as the ductility of the asphalt (expressed in centimeters).

Other test temperatures, such as 41 and 60°F (4 and 15°C), and loading speeds (1 cm/min) may be used depending on the specification requirements.

Solubility

The solubility test is a measure of the purity of the asphalt binder. The portion of the asphalt binder that is soluble in a specified solvent represents the active cementing constituents in an asphalt binder. Inert components—such as salts, free carbon, and nonorganic contaminants—are insoluble.

The solubility test was one of the first tests used to classify asphalt binders. At the time of the test development, carbon disulfide was used as the solvent. Due to its hazardous nature, the use of carbon disulfide as the solvent was discontinued and it was replaced by trichloroethylene. Other solvents, such as n-propyl bromide, may be used as allowed by the procedure.

Determining the solubility of an asphalt binder (Figure 4.24) involves dissolving approximately 2 grams of asphalt in 100 milliliters of solvent and then filtering the solution through a filter paper placed in a porcelain (Gooch) crucible with holes in the bottom to allow fluid to pass (Figure 4.25). The amount of material retained on the filter is determined by weighing and is expressed as a percentage of the original sample weight.

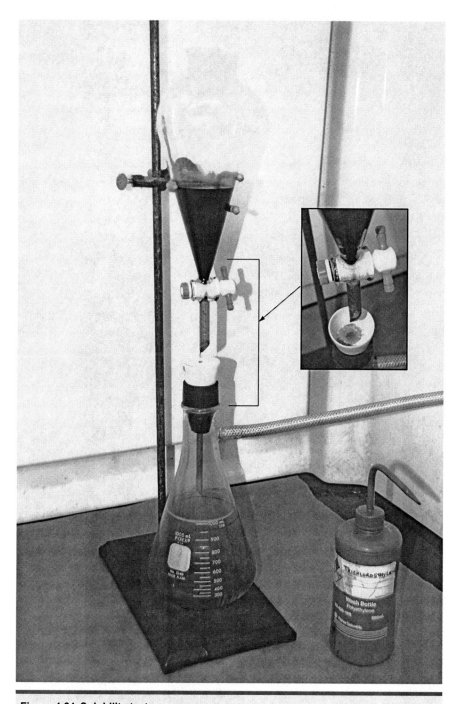

Figure 4.24 Solubility test

MS-26

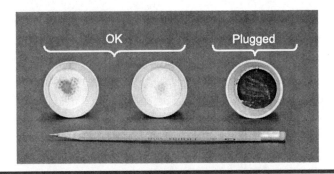

Figure 4.25 Solubility crucibles

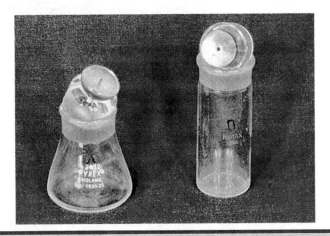

Figure 4.26 Specific gravity pycnometers

Specific gravity

The specific gravity of an asphalt binder can be used to convert between mass and volume. Knowing the asphalt binder specific gravity allows a user to calculate the mass of asphalt in a tank if the volume of asphalt in the tank is known. Specific gravity is also used in the calculation of some asphalt mixture properties.

The specific gravity test is performed by pouring heated asphalt binder into a glass pycnometer (Figure 4.26) which has been weighed and calibrated. After cooling, freshly boiled distilled or deionized water is added to the fill line. The pycnometer top is then lightly inserted and the pycnometer and sample are immersed in a water bath so that the sample acquires the appropriate test temperature. The pycnometer is removed from the water bath, wiped carefully, and weighed. Specific gravity is usually determined at either 60°F (15.6°C) or 77°F (25°C). It is important to note that, like other organic materials, the specific gravity changes with temperature.

Specifications for asphalt cements

In 1903, A. W. Dow presented a paper describing the use of various tests, including solubility, in a specification for ensuring the quality of asphalts. At that time, most of the asphalts in use were natural, not petroleum, asphalt binders containing significant amounts of mineral matter. Thus, the amount of insoluble organic and inorganic matter in a given sample was a way to identify the source of the asphalt binder. As asphalts from more than one source began to be used, specifications were written to attempt to control the quality of the asphalt binder.

Penetration-Graded specification

The Penetration-Graded specification (ASTM D 946) was first established in 1947 to attempt to control the quality of the asphalt binder using physical properties. Since the penetration value was the determining factor in identifying an asphalt binder's grade, the specification became known as the Penetration-Graded Specification.

In the Penetration Grading system, ASTM D 946 and AASHTO M 20, asphalt binders are classified into grades according to the penetration value measured at 77°F (25°C) of the unaged asphalt binder. Thus, an 85–100 (sometimes designated as 85/100) Pen asphalt binder would be an asphalt binder having a penetration value between 85 decimillimeters and 100 decimillimeters at 77°F (25°C). The softest (200–300) penetration grade asphalt is moderately firm at room temperature; at this temperature, gentle finger pressure indents the surface of the sample. The hardest (40–50) penetration grade asphalt is of a consistency that permits only a slight thumbprint under firm pressure when the material is at room temperature.

Advantages and shortcomings

While the Penetration-Graded Specification provides a means to evaluate the stiffness of an asphalt binder, it has some limitations. First, penetration is a test that generally measures a combination of viscous and elastic behavior in an asphalt binder. It is an empirical test in which the results must be correlated with pavement performance in order to understand the value of the result. The relationship between the test result and performance may not be very good, depending on the quality and quantity of the data used in the relationship.

Second, the Penetration-Graded Specification only measured asphalt stiffness at an intermediate temperature. How stiff the asphalt binder might be at other temperatures, as in the summer and winter, could only be inferred from the intermediate temperature. As conceptually illustrated

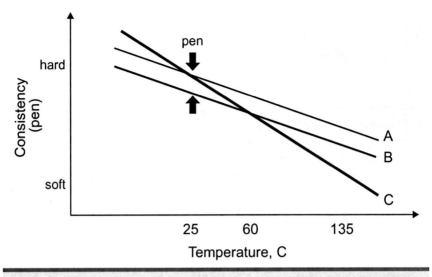

Figure 4.27 Penetration grading—asphalt binder properties

in Figure 4.27, the actual performance could be significantly different for asphalt binders with the same grade.

Finally, while short-term aging effects were captured through the use of the TFOT, there was no provision to conduct long-term aging of asphalt binders to simulate what happens to the asphalt stiffness after being in a pavement for a number of years.

Viscosity-graded specification

Beginning in the 1970s, viscosity began to be used more frequently than penetration in the grading of asphalt binders. ASTM D 3381 and AASHTO M 226 established specification criteria using absolute viscosity at 140°F (60°C) as the principal physical property for grading. In addition, viscosity at 275°F (135°C) is also usually specified. The purpose of the two criteria was to prescribe limiting values of consistency at two important temperatures. The 140°F (60°C) temperature was chosen as an approximation of the maximum temperature of asphalt pavement surfaces in service in the United States and elsewhere in the world. The 275°F (135°C) temperature was chosen as an approximation of the laydown temperature for hot mix asphalt. Thus, the specification addresses asphalt binder consistency during the laydown process of hot mix asphalt, when workability is an issue, and consistency at summer pavement temperatures, when stiffness is needed.

The AC system

In the Viscosity-Graded Specification, there are three tables that can be used to grade asphalt binders. Tables 1 and 2 use the AC (Asphalt Cement) system consisting of AC-2.5, AC-5, AC-10, AC-20, AC-40, and sometimes

AC-30 grades. The numerical values in the grade designation denote the viscosity of the unaged asphalt binder in hundreds of poises at 140°F (60°C). Thus, an AC-10 would be expected to have a viscosity of approximately 1000 poises at 140°F (60°C). In each grade, the allowable tolerance on viscosity is ± 20 percent. For example, an AC-10 grade would include asphalt binders with viscosity of 800–1200 poises at 140°F (60°C).

Table 1 and Table 2 use the same tests but have different criteria for a couple of tests. For a given grade, the viscosity at 140°F (60°C) will have the same range in Table 1 and Table 2. However, the required minimum viscosity at 275°F (135°C) and penetration at 77°F (25°C) is lower for a Table 1 asphalt binder than a Table 2 asphalt binder. As stated in Note 2 of the specification (ASTM D 3381), this implies that Table 2 asphalt binders are less temperature susceptible than Table 1 asphalt binders of the same grade.

The AR system

Table 3 in the Viscosity-Graded Specification uses the AR (Aged Residue) system consisting of AR-1000, AR-2000, AR-4000, AR-8000, and AR-16000 grades. Like the AC system, the numerical values in the grade designation denote the viscosity in poises at 140°F (60°C), but with the viscosity being measured after the asphalt has been subjected to the rolling thin film oven test (RTFOT). In each grade, the allowable tolerance on viscosity is ± 25 percent. Thus an AR-4000 grade would include asphalt binders with viscosity, after the RTFOT, of 3000–5000 poises at 140°F (60°C).

Advantages and shortcomings

While the Viscosity-Graded Specification represents an improvement over the Penetration-Graded Specification, it also has some of the same limitations. Viscosity measurements at 275°F (135°C) allow the user to determine the stiffness of the asphalt binder at typical production temperatures for hotmix asphalt mixtures. Viscosity measurements at 140°F (60°C) allow the user to determine the stiffness of the asphalt binder at typical high pavement temperatures. Despite the addition of intermediate temperature measurements (such as penetration) to some specifications, the stiffness of the asphalt binder at cold temperatures could only be inferred from the intermediate temperature data. As in Figure 4.27, the data in Figure 4.28 illustrate that the actual performance could be significantly different for asphalt binders with the same grade.

As with the Penetration-Graded Specification, short-term aging effects are captured through the use of the TFOT or RTFOT, but there is no provision to conduct long-term aging of asphalt binders to simulate what happens to the asphalt stiffness after being in a pavement for a number of years.

Although viscosity is a fundamental property, the viscosity test is only useful in measuring the viscous component of an asphalt binder. This is not an issue for conventional, refined asphalt binders but becomes a greater issue when using asphalt binders produced using polymer modification. For these asphalt binders, the elastic component of the asphalt binder is not properly considered when using the viscosity test. As such, the viscosity

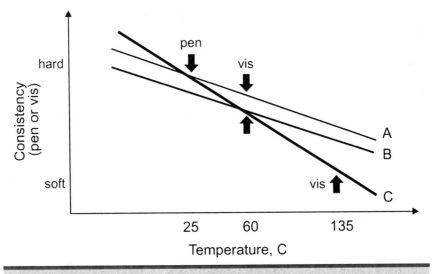

Figure 4.28 Viscosity grading—asphalt binder properties

of a modified-asphalt binder may not provide sufficient information about the asphalt binder's physical properties to properly judge its performance potential.

In addition to the principal tests (penetration or viscosity) for defining grade, several other tests are required to determine specific properties for these products. For general reference, the tests are listed in Table 4.1, along with the appropriate AASHTO or ASTM test method and indications of which specification requires the test. Readers should always refer to the current AASHTO or ASTM specification.

As a final note, since viscosity is a fundamental material property and penetration is not and they are measured at two different temperatures, there is no universal relationship between the two properties. The relationship between penetration and viscosity varies for different asphalts produced from different crude sources. However, there is a general trend where higher penetration values are associated with lower viscosity values and vice versa. This is illustrated in Figure 4.29 where the penetration values are correlated with viscosity values for the 48 asphalt binders in the Materials Reference Laboratory (MRL) of the Strategic Highway Research Program.

Table 4.1
Asphalt cement specification tests

	Test	Test Method		Specification[3]	
		AASHTO	ASTM	Viscosity	Penetration
Unaged Asphalt	Viscosity @ 140°F (60°C)	T 202	D 2171	x	
	Viscosity @ 275°F (135°C)	T 201	D 2170	x	
	Penetration @ 77°F (25°C)	T 49	D 5	x	x
	Flash Point	T 48	D 92	x	x
	Solubility in TCE	T 44	D 2042	x	x
	Ductility @ 77°F (25°C)	T 5	D 113		x
	Spot Test[1]	T 102	n/a	x	x
Aged Residue	Mass Loss[2]	T 179 / T 240	D 1754 / D 2872	x / x	x
	Viscosity @ 140°F (60°C)	T 202	D 2171	x	
	Viscosity @ 275°F (135°C)	T 201	D 2170	x	
	Penetration @ 77°F (25°C)	T 49	D 5	x	x
	Ductility @ 77°F (25°C)	T 5	D 113	x	x

[1] Optional test in AASHTO specification.
[2] Mass Loss is conducted by either TFO or RTFO (TFO listed first). Residue is obtained after aging.
[3] Viscosity specifications are AASHTO M 226 and ASTM D 3381. Penetration specifications are AASHTO M 20 and ASTM D 946. Note that the viscosity specification includes three grading tables for which test requirements may differ from what is shown in this table.

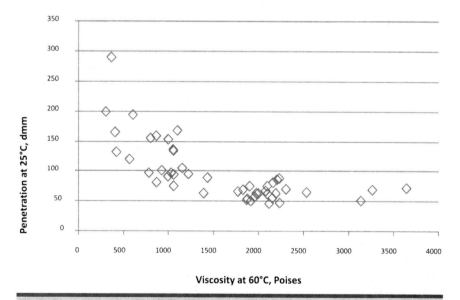

Figure 4.29 Pen-Vis correlation for MRL asphalt binders

PG (Performance Graded) asphalt binder tests

Recognizing some of the shortcomings in the traditional specification systems, industry began to work towards the development of a specification in the late 1980s and early 1990s that is performance related, based on fundamental physical properties. The Performance Graded Asphalt Binder Specification was formed around both new and existing physical property tests.

Rotational viscometer

The rotational viscosity test is used to determine the flow characteristics of an asphalt binder at a high temperature to provide some assurance that it can be pumped and handled at the hot-mixing facility. A rotational coaxial cylinder viscometer, as shown in Figure 4.30 and described in AASHTO T 316, is necessary to evaluate the various types of asphalt binders. Unlike capillary tube viscometers, rotational viscometers have larger clearances between the components, allowing for better applicability to both modified and unmodified asphalts.

When a rotational viscometer is used, viscosity is determined by measuring the torque required to maintain a constant rotational speed of a cylindrical spindle that is submerged in an asphalt binder sample at a constant temperature (Figure 4.31). This torque is directly related to the asphalt binder viscosity, which is calculated automatically by the viscometer. Because this binder viscosity is used to ensure that the asphalt is fluid enough at normal operating temperatures to pump and mix with aggregate, it is measured on original (unaged) or "tank" asphalt. The viscometer can also be used to develop temperature-viscosity charts for estimating mixing and compaction temperatures for use in mixture design.

In the rotational viscosity test, a sample of asphalt binder is weighed into a preheated sample chamber. The amount of asphalt used is typically 8 to 11 grams and varies with the size of spindle. The sample chamber containing the binder sample is placed in the preheated thermo-container (Figure 4.32) operating at the proper test temperature (usually 135°C). The preheated spindle is lowered into the sample, and the binder is ready to test when the temperature stabilizes, usually within 30 minutes.

At the end of this time, the rotational viscosity test is started by activating the motor so that the spindle turns at a prescribed rate—20 rpm for specification testing. As the spindle rotates, the torque in the torsional spring is measured and converted to viscosity. A waiting period of 15 minutes is needed to reach a uniform sample temperature. During this period with the spindle rotating, the viscosity reading and the percentage of torque can be observed on the digital display. If the percentage of torque is less than 2 percent

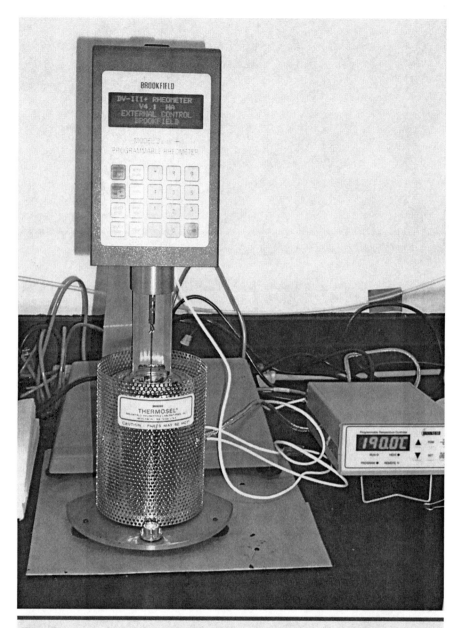

Figure 4.30 RV

or greater than 98 percent, then a different-size spindle may be required. Alternatively, the rotational speed can be increased or decreased to change the shear rate and thereby change the percentage of torque. For a Newtonian fluid, as are most unmodified asphalt binders at high temperature, changing the shear rate (rotational speed) should not have any effect on the measured viscosity. As the temperature equalizes, the viscosity reading will stabilize

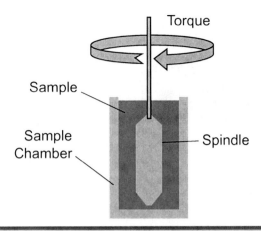

Figure 4.31 RV operation

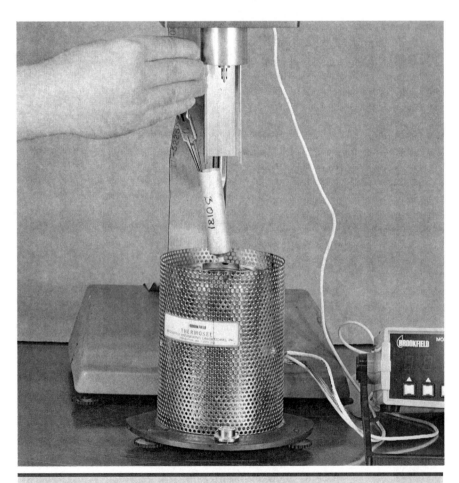

Figure 4.32 Inserting sample chamber into thermal chamber

and test results are recorded. Three viscosity readings are recorded at 1-minute intervals, and the average of the three readings is reported.

Users are cautioned to check the units reported by the rotational viscometer because the digital output of some rotational viscometers is in units of centipoise (cP). The conversion of centipoises to Pascal-seconds, or Pa•s, is given by the equation 1000 cP = 1 Pa•s. Therefore, to obtain viscosity in Pa•s, rotational viscosity in cP is divided by 1000.

Finally, besides testing at 135°C for specification purposes, it may be desirable to determine the asphalt binder viscosity at additional temperatures. For instance, many asphalt technologists use equiviscous temperatures for mixing and compaction during mix design. To determine these temperatures, a viscosity-temperature relationship for the asphalt binder is needed. This means that the technologist must determine the viscosity at an additional—usually higher—temperature, such as 163°C. The rotational viscosity test is conducted as before but at the new test temperature.

Dynamic Shear Rheometer (DSR)

The dynamic shear rheometer is used to test asphalt binders and measure their rheological properties, including complex shear modulus (G^*, pronounced "G-star") and phase angle (δ, the Greek letter "delta"), at a broad range of temperatures. These parameters can be used to characterize both viscous and elastic behavior of asphalt binders. The complex shear modulus, G^*, is a measure of the total resistance of a material to deformation when exposed to a sinusoidal shear stress load. G^* consists of both elastic (recoverable) and viscous (nonrecoverable) components. The phase angle, δ, is an indicator of the relative amounts of viscous and elastic components. The values of G^* and δ for asphalt binders are highly dependent on the test temperature and frequency of loading. At high temperatures (and slow loading frequencies), asphalt binders behave like viscous fluids with little capacity for recovering or rebounding. In this case, the asphalt binder could be represented by the vertical axis (viscous component only) in Figure 4.33; there would be no elastic component of G^*, since $\delta = 90°$. At very low temperatures, asphalts behave like elastic solids, which rebound from deformation completely. This condition is represented by the horizontal axis (elastic component only) in Figure 4.33. In this case, there is no viscous component of G^*, since $\delta = 0°$.

Under normal pavement temperatures and traffic loading, asphalt binders act with the characteristics of both viscous liquids and elastic solids and are said to be "viscoelastic" materials. By measuring G^* and δ, the DSR provides an indication of the behavior of asphalt at a range of pavement service temperatures. The vector arrows in Figure 4.33, G^*_1 and G^*_2, represent the complex modulus values of Binders 1 and 2. When these asphalt binders are loaded, part of their deformation is elastic *(E)* and part is viscous *(V)*. Even though both asphalt binders in Figure 4.33 are viscoelastic and have the same G^*, Binder 2 is more elastic than Binder 1, because

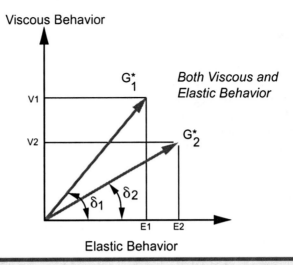

Figure 4.33 Graphic representation of shear modulus and phase angle

of its smaller δ. Because Binder 2 has a larger elastic component, it will recover much more deformation from an applied load than Binder 1. This example clearly shows that G^* alone cannot describe asphalt behavior; the value of δ is also needed.

The operation of a DSR is fairly basic. A sample of asphalt binder is sandwiched between two parallel plates, one that is fixed and one that oscillates (Figure 4.34). As the plate oscillates, the centerline of the plate at point A (indicated by the dark vertical line) moves to point B. From point B, the plate centerline moves back and passes point A to point C. From point C, the plate centerline moves back to point A. This oscillation is one cycle and is repeated for a specified number of cycles. The number of cycles completed in one second is the loading frequency. Typically, DSR tests that are run for specification purposes are performed at a frequency of approximately 1.59 hertz (1.59 cycles per second)—equivalent to 10 radians per second. Stress and strain are measured during each loading cycle and used to calculate complex shear modulus and phase angle.

To conduct testing, the DSR equipment (Figure 4.35) is first initialized to operate with the proper parallel plate geometry and gap. The plate diameter and gap thickness used depend on the aged state of the asphalt being tested. Original (unaged) and short-term-aged binders are tested with a 25-millimeter diameter plate and a 1000-micron (1-millimeter) gap; long-term-aged binder is tested with an 8-millimeter diameter plate and a 2000-micron (2-millimeter) gap. Before mounting the specimen, the gap is first set to zero at the test temperature and then set at the desired value (1000 or 2000 microns) plus an extra gap, usually 50 microns, to allow for the creation of a slight bulge in the test specimen.

To prepare the test specimen, the asphalt binder is heated until fluid, and is poured either directly onto one of the two parallel plates or into a

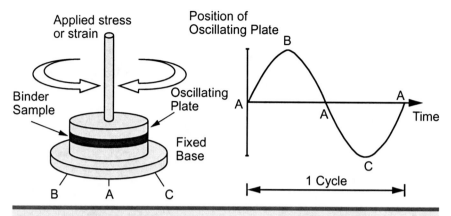

Figure 4.34 DSR operation

Figure 4.35 DSR

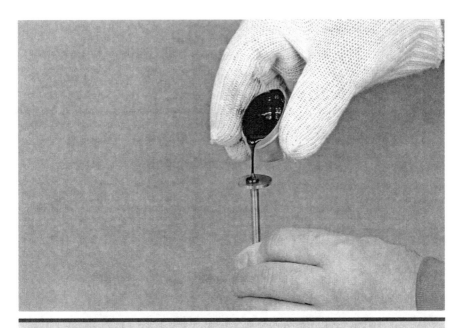

Figure 4.36 Direct pour

silicone mold with the appropriate diameter and thickness for testing. In some rheometers, the top parallel plate can be removed without affecting the 0 gap setting. In this case, the asphalt binder sample can either be poured directly onto the plate (Figure 4.36) or transferred to the plate using a spatula or glass rod. The asphalt binder can also be poured directly onto the bottom parallel plate. Some technologists prefer the direct pour method since the asphalt binder is heated when it is applied to the plate, thereby promoting adhesion between the plates and the asphalt binder sample.

Alternatively, some technologists use a silicone rubber mold (Figure 4.37) to prepare the asphalt binder sample. The advantage of the silicone mold is that it, when properly filled, produces an asphalt binder test sample that is the appropriate size for the DSR and minimizes excess material, which can lengthen the trimming process. Once an asphalt binder sample has been poured into the silicone mold, it should be allowed to cool at room temperature for a short amount of time (30 minutes is generally sufficient) until the sample pellet can be removed from the mold. At that time, the sample can be directly applied to either the top or bottom parallel plate (Figures 4.38 and 4.39). The asphalt binder sample should not be touched during this process as contaminants (oils, dirt) from the hands could impact the test results. For soft asphalt binders, technologists may need to place the silicone mold and asphalt binder sample in a freezer for a short time (usually 5–15 minutes is sufficient) to allow the sample to be demolded.

Figure 4.37 Silicone mold

Figure 4.38 Transfer to top plate

MS-26

71

Figure 4.39 Transfer to bottom plate

Silicone molds are an expedient means of producing DSR samples, but the technologist needs to remember that asphalt binders can become stiffer over time at room temperature—a process called steric hardening (discussed in Chapter 3). Samples should be tested as quickly as reasonably possible and not be allowed to sit at room temperature for an extended period of time.

After the asphalt binder is loaded, the parallel plates are brought together until achieving the appropriate gap setting (either 1,000 or 2,000 microns) plus the extra gap amount. With the asphalt binder sample sandwiched between the plates, the sample is trimmed so that the asphalt binder is flush with the edges of the parallel plates (Figure 4.40). This trimming is needed to control the diameter of the test specimen and enable an accurate calculation of G^*. After trimming, the extra gap is eliminated so that the final gap, or specimen thickness, is exactly at the desired value. The specimen will bulge slightly (Figures 4.41 and 4.42).

Because asphalt binder properties are temperature dependent, DSRs must have a precise means of controlling the sample temperature. This control is accomplished using either a circulating-fluid bath or a forced-air oven. In either case, the temperature must be controlled so that the sample temperature is uniform and varies by no more than 0.1°C across the gap.

Once the asphalt binder specimen is loaded and trimmed, the temperature of the test specimen needs time to equilibrate. While a minimum equilibration time of 10 minutes is commonly used, the actual temperature equilibration time is equipment dependent and should be checked using a calibration specimen with very accurate temperature-sensing capabilities.

Figure 4.40 Trimming

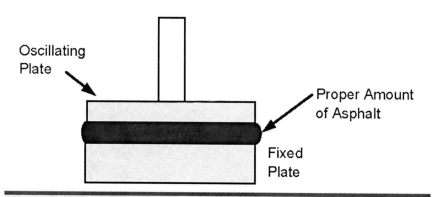

Figure 4.41 Asphalt sample configuration

Before testing, the operator sets the approximate value of shear strain to be achieved during the test. Desired strain values vary from about 1 percent to 12 percent and depend on the aged state of the binder being tested. Original (unaged) binders and short-term-aged binders are tested at strain values of approximately 10 to 12 percent. Long-term-aged binders are tested at strain values of about 1 percent. In all cases, strain values must be small enough that the response of the binder (G^*) remains in the linear viscoelastic (LVE) range. As long as the test remains within the LVE range of the material, G^* will be virtually unaffected by changes in strain level.

Figure 4.42 Sample after trimming

Once the test begins, the equipment software controls the test and provides the proper sinusoidal loading while measuring stress and strain responses. The sample is first conditioned by loading for 10 cycles. During this conditioning period, the DSR measures the stress required to achieve the desired shear strain and then precisely maintains this stress during the test. After the 10 conditioning cycles, 10 additional cycles are applied to obtain test data. Software automatically computes and reports G^* and δ, using the relationship between the applied stress and the resulting strain.

The complex shear modulus, G^*, is the ratio of total shear stress ($\tau_{max} - \tau_{min}$) to total shear strain ($\gamma_{max} - \gamma_{min}$). The time lag between the applied stress and the resulting strain (or applied strain and resulting stress) is related to the phase angle, δ. For a perfectly elastic material, an applied load causes an immediate response; thus, the time lag or phase angle is zero. A viscous material (such as hot asphalt at mixing temperatures) has a relatively large time lag between load and response; in this case, the angle approaches 90 degrees. Because asphalt binders are viscoelastic at normal pavement temperatures, they behave somewhere between the 2 extremes. These responses are illustrated in Figures 4.43 and 4.44.

The formulas used by the rheometer software to calculate τ_{max} and γ_{max} are shown in Figure 4.45. The formulas are not shown so that the user can actually calculate G^*—this is accomplished by the DSR operating software—but to illustrate the importance of the specimen geometry. In the determination of G^*, the radius of the specimen is raised to the fourth power. Thus, errors in trimming the test specimen can rapidly impact the calculated G^* value.

Once testing is completed, a sample report is generated (Figure 4.46) and the appropriate specification value is compared to the criterion.

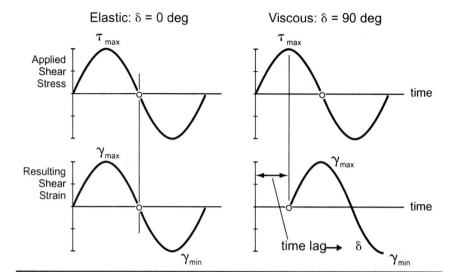

Figure 4.43 Stress-strain output

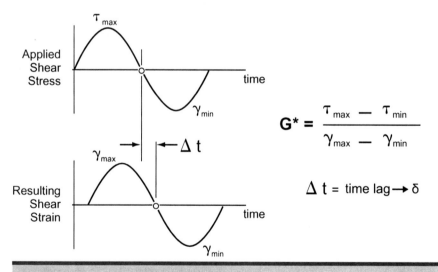

Figure 4.44 Stress-strain output of VE material

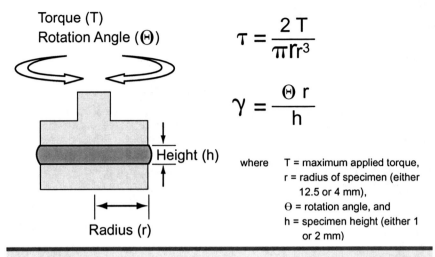

Figure 4.45 DSR calculations

```
AASHTO T315 Original Binder   RHEOPLUS/32 V3.40 21002706-33025
===============================================================

MEASURING DATA:
No.  Time    Temp.   Freq.    Deform.   |G*|    delta
[1]  [s]     [°C]    [rad/s]  [%]       [kPa]   [°]
1    428.0   64.0    10.0     12.00     1.34    88.8
2    433.8   64.0    10.0     12.00     1.34    88.8
3    439.7   64.0    10.0     12.00     1.34    88.8
4    445.5   64.0    10.0     12.00     1.34    88.8
5    451.3   64.0    10.0     12.00     1.34    88.8
6    457.2   64.0    10.0     12.00     1.34    88.8
7    463.0   64.0    10.0     12.00     1.34    88.8
8    468.9   64.0    10.0     12.00     1.34    88.8
9    474.7   64.0    10.0     12.00     1.34    88.8
10   480.5   64.0    10.0     12.00     1.34    88.8

TEST RESULTS:
Mean frequency                      :    10.0 rad/s = 1.59 Hz
Mean strain amplitude               :    12.00 %
Mean stress amplitude               :    0.161 kPa
Mean phase angle                    :    88.8 °
Mean complex modulus |G*|           :    1.34 kPa
Mean temperature lower plate        :    64.0 °C
Mean temperature sample             :    64.0 °C

EVALUATION RESULTS:
Number of data points               :    10
|G*|/sin(delta)                     :    1.34 kPa
    Standard deviation              :    0.000454 kPa
    Median                          :    1.34 kPa
    Confidence Interval (95%)       :    1.34 ... 1.34 kPa
AASHTO T315 ORIG. BINDER PERF. CRIT.:    |G*|/sin(delta) >= 1.00 kPa

===============================================================
            This sample is found to  P A S S  at 64.0 °C
===============================================================
```

Figure 4.46 Sample report

```
AASHTO T315 Original Binder   RHEOPLUS/32 V3.40 21002706-33025

ANALYSIS PARAMETERS:
Number of data points              : 10
Testing boundary                   : 1.00 kPa
Temperature offset                 : 0 °C
Target temperature                 : Automatic
Show parameter settings            : On
Show result table                  : On
Show PG suggestion                 : Off
Suppress temp. detection warning   : Off
Fill with blank lines              : Off

TEST PARAMETERS:
File name                          : C:\RHEOPLUS\USER\000963o.orx
Data series                        : Verification: ORIGINAL BINDER
Used measuring points              : I 2, P 11 ... I 2, P 20
Date of test                       : 6/3/2010
Time of test                       : 8:51:26
Operator                           : ajp
Sample                             : 000963o
Remark                             :
Measuring device                   : SmartPave101 SN80064763; FW3.30; Slot4; Adj9d
Temperature control unit           : TU1=P-PTD200+H-PTD120-SN80064756-841443
Measuring system                   : PP25/PE-SN13149
    Diameter                       :   25.00 mm
    Gap                            :    1.000 mm

MEASURING DATA:
No.  Time   Temp.   Freq.    Deform.   |G*|     delta
[1]  [s]    [°C]    [rad/s]  [%]       [kPa]    [°]
1    428.0  64.0    10.0     12.00     1.34     88.8
2    433.8  64.0    10.0     12.00     1.34     88.8
3    439.7  64.0    10.0     12.00     1.34     88.8
4    445.5  64.0    10.0     12.00     1.34     88.8
5    451.3  64.0    10.0     12.00     1.34     88.8
6    457.2  64.0    10.0     12.00     1.34     88.8
7    463.0  64.0    10.0     12.00     1.34     88.8
8    468.9  64.0    10.0     12.00     1.34     88.8
9    474.7  64.0    10.0     12.00     1.34     88.8
10   480.5  64.0    10.0     12.00     1.34     88.8

TEST RESULTS:
Mean frequency                     :   10.0 rad/s = 1.59 Hz
Mean strain amplitude              :   12.00 %
Mean stress amplitude              :    0.161 kPa
Mean phase angle                   :   88.8 °
Mean complex modulus |G*|          :    1.34 kPa
Mean temperature lower plate       :   64.0 °C
Mean temperature sample            :   64.0 °C

EVALUATION RESULTS:
Number of data points              :   10
|G*|/sin(delta)                    :    1.34 kPa
    Standard deviation             :    0.000454 kPa
    Median                         :    1.34 kPa
    Confidence Interval (95%)      :    1.34 ... 1.34 kPa
AASHTO T315 ORIG. BINDER PERF. CRIT.: |G*|/sin(delta) >= 1.00 kPa

            This sample is found to  P A S S  at 64.0 °C
```

Figure 4.46 Sample report

Bending Beam Rheometer (BBR)

Asphalt binders at low temperatures are too stiff to reliably measure rheological properties using the parallel-plate geometry of a conventional DSR. Consequently, SHRP researchers developed the bending beam rheometer (BBR) to accurately evaluate binder properties at low pavement temperatures. Used together, the DSR and BBR tests provide stiffness behavior of asphalt binders over a wide range of temperatures.

The BBR is used to measure how much a binder deflects or creeps under a constant load at a constant temperature. BBR test temperatures are related to a pavement's lowest service temperature, when the asphalt binder acts more like an elastic solid than a viscous fluid. Furthermore, the test is performed on binders that have been aged to simulate hot mixing in a mixing facility and some in-service aging.

The BBR gets its name from the test specimen geometry and loading method used during testing. The key elements of the BBR (Figure 4.47) are a loading frame, temperature-controlled fluid bath, and the computer-control, data-acquisition system. A blunt-nosed shaft applies a load to the midpoint of a simply supported asphalt binder beam. A load cell is mounted on the loading shaft, which is enclosed in an air bearing to eliminate friction during loading. Loads are applied by pneumatic pressure. A transducer on the loading shaft monitors deflection (Figure 4.48).

The BBR procedure uses beam theory to calculate the stiffness of an asphalt beam sample under a creep load. By applying a constant load to the asphalt beam and measuring the center deflection of the beam throughout the four-minute test procedure, the creep stiffness *(S)* and creep rate *(m-value)* can be calculated. *Creep stiffness* is the resistance of the asphalt binder to creep loading, and the *m-value* is the change in asphalt stiffness with time during loading.

The controlled temperature bath contains a fluid that will not freeze at the test temperatures used (potentially as low as $-36°C$). Fluids such as

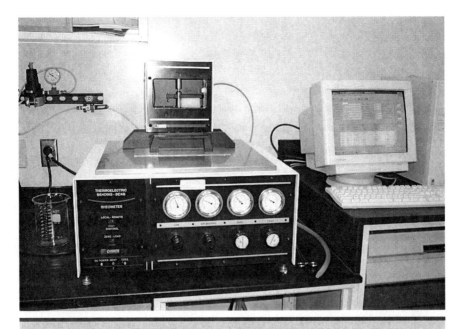

Figure 4.47 BBR

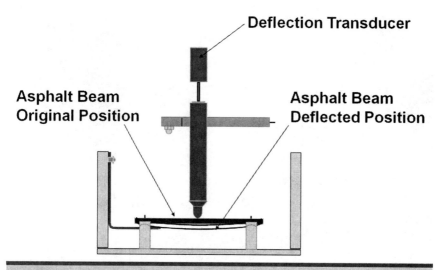

Figure 4.48 BBR operation

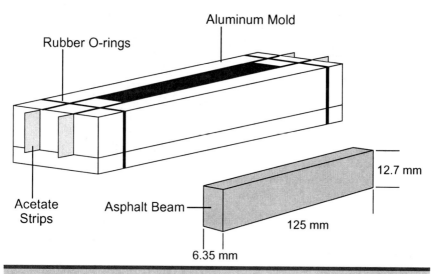

Figure 4.49 BBR mold

methanol, ethanol, ethylene glycol, water, and combinations of these are commonly used. As with the DSR, tight temperature control is needed (± 0.1°C).

Asphalt binder test specimens are prepared using a rectangular aluminum mold (Figure 4.49). The mold assembly consists of an aluminum base, side and end plates, acetate base and side strips, and rubber O-rings. Before the mold is assembled, the inside surfaces of the 2 side plates and

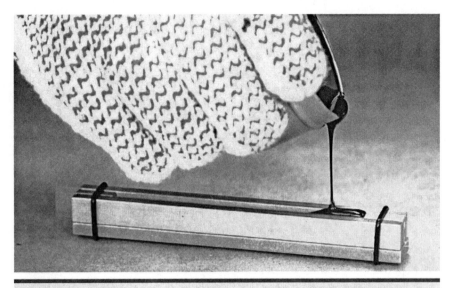

Figure 4.50 Pouring

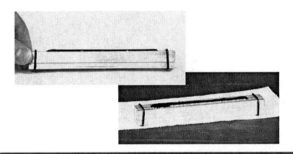

Figure 4.51 Properly poured test specimen

the base plate are lightly coated with a petroleum-based jelly. The three acetate strips are then placed against the greased faces. The end pieces are treated with a release agent composed of glycerin and talc. The mold is then assembled and held together with 2 rubber O-rings. The asphalt beam specimens are formed by heating the binder until fluid and filling the mold in a continuous motion from one end to the other (Figures 4.50 and 4.51).

After a cooling period of 45 to 60 minutes, excess asphalt is trimmed from the upper surface using a hot putty knife (Figure 4.52). The asphalt specimen remains in the mold at room temperature for a short time and is removed from the mold only when the conditioning procedure is ready to begin. As discussed earlier, steric hardening can cause the asphalt binder sample to stiffen over time at room temperature, which could affect the test results.

So that the asphalt binder test specimen can be removed from the mold, the assembly is cooled for 5 to 10 minutes in an ice bath or freezer. The mold is

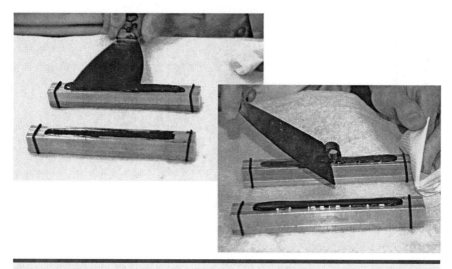

Figure 4.52 Trimming

then rapidly disassembled and the asphalt binder beam is transferred immediately to the BBR bath for conditioning at the test temperature for 60 minutes. As discussed earlier, asphalt binder samples can experience an increase in stiffness over time when held at low temperatures. This phenomenon of physical hardening (discussed in Chapter 3) is similar to steric hardening and could negatively impact the quality of the test results. For this reason, technologists must strictly adhere to the conditioning time (60 ± 5 minutes).

After the thermal conditioning period in the bath, the asphalt beam is loaded in the test frame and subjected to a constant creep load of 980 MilliNewtons for 240 seconds. Load and beam deflection are continuously measured during the test and are used to calculate creep stiffness *(S)* and creep rate *(m-value)* at specified load times—8, 15, 30, 60, 120, and 240 seconds—using beam theory. Figure 4.53 shows a sample graph and the procedure used to obtain the creep rate *(m-value)* at each load time.

To relate to field performance, researchers determined that the value of creep stiffness should be determined when the asphalt binder has been loaded for 2 hours at the minimum pavement design temperature. However, when using the concept of time-temperature superposition, SHRP researchers also determined that by increasing the test temperature 10°C, an equivalent creep stiffness can be obtained after only 60 seconds of loading. The obvious benefit is that a test result can be obtained in a much shorter testing time (60 seconds instead of 2 hours).

The second parameter determined from the results of the bending beam test is the *m-value*. The *m-value* represents the rate of change in the creep stiffness, $S(t)$, versus time. This value is also calculated automatically by the bending beam software. To obtain *m-value*, the stiffness is calculated at several loading times (Figure 4.53). The *m-value* is the slope of the log stiffness versus log time curve at any time, t.

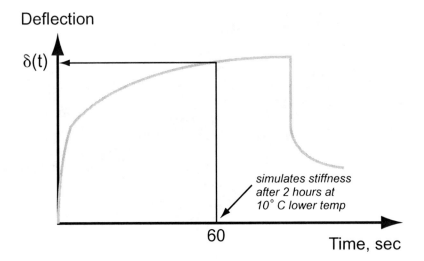

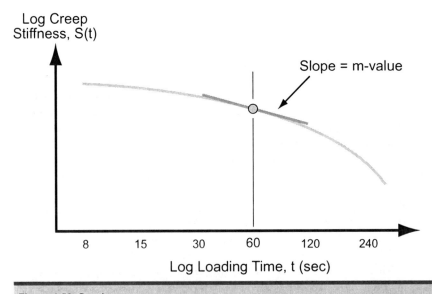

Figure 4.53 Graphs

Computer-generated output for the bending beam test includes all the required reporting items (Figure 4.54). The output format generally includes plots of deflection and load versus time, actual load and deflection values at various times, test parameters, and operator information. Technologists should compare the creep stiffness, S(60), and *m-value*, m(60), at 60 seconds loading at the specified test temperature to the appropriate criteria.

CANNON® Instrument Company, USA   1.23   06/04/2010 05:01:57 PM

Project :
Operator : jal
Specimen : 000963-1
Test Time : 04:56:47 PM
Test Date : 06/04/2010
File Name : 10060413
BBR ID : bbr#2

Target Temp (°C) : -12.0
Min. Temp (°C) : -12.0
Max. Temp (°C) : -12.0
Temp Cal Date : 06/04/2010
Soak Time (min) : 55.0
Beam Width (mm) : 12.70
Thickness (mm) : 6.35

Conf Test (GPa) : 218
Conf Date : 06/04/2010
Force Const (mN/bit) : 0.149
Defl Const (μm/bit) : 0.152
Cmpl (μm/N) : 7.45
Cal Date : 06/04/2010
Software Version : BBRw 1.23

t Time (s)	P Force (mN)	d Deflection (mm)	Measured Stiffness (MPa)	Estimated Stiffness (MPa)	Difference (%)	m-value
8.0	980	0.205	385	386	0.260	0.247
15.0	981	0.241	328	328	0.000	0.271
30.0	981	0.294	269	269	0.000	0.298
60.0	981	0.365	217	217	0.000	0.325
120.0	981	0.461	172	172	0.000	0.352
240.0	981	0.594	133	133	0.000	0.379

A = 2.77   B = -0.166   C = -0.0448 R^2 = 0.999997

Force (t=0.0s) =   34 mN   Deflection (t=0.0s) = 0.000 mm
Force (t=0.5s) = 948 mN   Deflection (t=0.5s) = 0.107 mm

Max Force Deviation (t=0.5 - 5.0s) =   -33, +0 mN
Max Force Deviation (t=5.0 - 240.0s) =   -1, +1 mN

Average Force (t=0.5 - 240.0s) = 981 mN
Maximum Force (t=0.5 - 240.0s) = 982 mN
Minimum Force (t=0.5 - 240.0s) = 948 mN

CANNON® Instrument Company, USA   1.1   01/15/2003 10:42:00

Project :
Operator : se
Specimen : 02-322 #1
Test Time : 10:35:56
Test Date : 01/15/2003
File Name : 03011503
BBR ID : Al Cannon BBR

Target Temp (°C) : -12.0
Min. Temp (°C) : -12.1
Max. Temp (°C) : -12.0
Temp Cal Date : 10/22/2002
Soak Time (min) : 60.0
Beam Width (mm) : 12.70
Thickness (mm) : 6.35

Conf Test (GPa) : 199
Conf Date : 01/15/2003
Force Const (mN/bit) : 0.149
Defl Const (μm/bit) : 0.159
Cmpl (μm/N) : 5.15
Cal Date : 01/15/2003
Software Version : BBRw 1.1

t Time (s)	P Force (mN)	d Deflection (mm)	Measured Stiffness (MPa)	Estimated Stiffness (MPa)	Difference (%)	m-value
8	991.3	0.214	373	373	0.000	0.249
15	990.6	0.252	317	317	0.000	0.272
30	990.1	0.307	260	260	0.000	0.296
60	990.4	0.379	211	210	-0.474	0.321
120	991.6	0.479	167	167	0.000	0.346
240	993.8	0.615	130	130	0.000	0.370

A = 2.76   B = -0.175   C = -0.041 R^2 = 0.999990

Force (t=0.0s) =   31.8 mN   Deflection (t=0.0s) = 0.000 mm
Force (t=0.5s) = 984.5 mN   Deflection (t=0.5s) = 0.114 mm

Figure 4.54 Sample report

Direct Tension Tester (DTT)

Numerous past studies have shown a strong relationship between the stiffness of conventional (unmodified) asphalt binders and the amount of stretching they undergo before breaking. Asphalts that undergo considerable stretching before failure are called ductile; those that break without much stretching are called brittle. It is important that an asphalt binder be capable of at least minimal elongation. Typically, stiffer asphalts are more brittle and softer asphalts are more ductile.

Creep stiffness as measured by the BBR is not adequate to completely characterize the capacity of asphalts to stretch before breaking. For example, some binders exhibit high creep stiffness but can also stretch farther before breaking. To address this issue, SHRP researchers developed a test to evaluate the fracture properties of asphalt binders at low temperatures.

The direct tension test measures the amount of binder stress and strain at failure at very low temperatures. It is performed at temperatures where asphalt binders generally exhibit brittle behavior (0°C to −36°C). Furthermore, the test is performed on asphalt binders that have been subjected to long-term (PAV) aging to simulate several years of in-service aging.

Although the direct tension test concept is simple, the equipment (Figure 4.55) used requires complex features to accurately measure the very small strains involved in the test. The direct tension tester (DTT) consists of three components: an electromechanical test device to apply elongation, a transducer system to measure load on the specimen, and an environmental control system (usually in the form of a controlled temperature

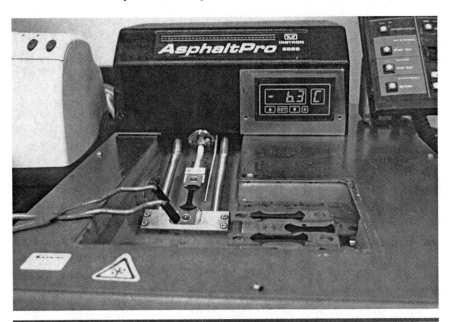

Figure 4.55 DTT

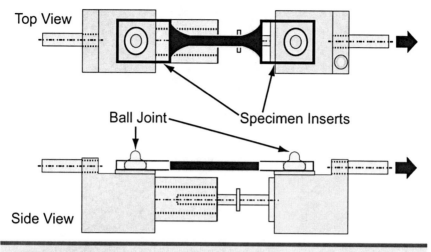

Figure 4.56 DTT loading

liquid bath). A key feature of the testing machine is the gripping system used to attach specimens to the alignment rods that apply the tensile load (Figure 4.56). The grips have a ball-joint connection that ensures no bending within the binder test specimen.

In the direct tension test, a small asphalt specimen shaped like a dog bone (Figure 4.57) is pulled at a slow, constant rate until it fails. The elongation at failure is used to calculate the failure strain, which is an indication of whether the binder will behave in a brittle or ductile manner at the low test temperature (Figure 4.58). The failure strain (ε_f) is calculated as the change in length divided by the original length, expressed as a percentage. In the direct tension test, failure is defined as the load when the stress reaches its maximum value, not necessarily the load when the specimen breaks. Failure stress (σ_f) is the failure load divided by the original cross section of the specimen.

In preparation for running the direct tension test, asphalt binder test specimens are formed in aluminum molds with end inserts, made from a phenolic resin and containing a hole inset with a stainless steel ring (Figure 4.59). As in the BBR test, heated asphalt binder is poured into the mold, filling it in a continuous motion from one end to the other (Figure 4.60). Six individual specimens are formed to produce an average test result.

After the specimens are poured (Figure 4.61), they are allowed to cool at ambient temperature for 30–60 minutes before trimming (Figure 4.62). After trimming, the specimen is again allowed to cool for 10–15 minutes before it is carefully removed from the mold assembly. As in the BBR test, the test specimens are then subjected to temperature conditioning in the fluid bath at the specified test temperature for 60 minutes before testing. Once again, this temperature conditioning period is tightly controlled to minimize physical hardening effects.

The environmental control system includes an environmental chamber (or liquid bath) and a mechanical refrigeration unit capable of producing and

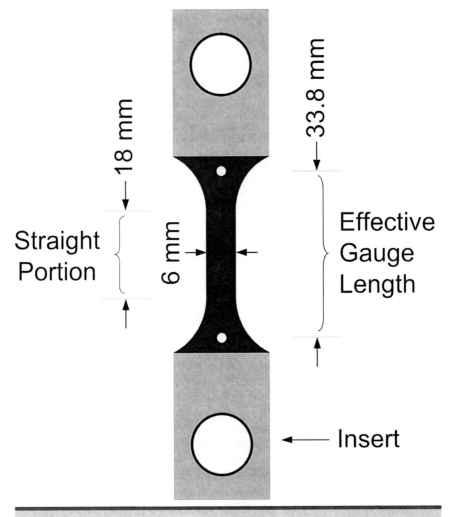

Figure 4.57 DTT specimen

precisely maintaining temperatures as low as −36°C. In the case of an environmental chamber, air is used as the cooling medium. If a liquid bath is used, a solution of potassium acetate and water is used as the cooling medium.

At the end of the conditioning period (60±5 minutes), each of the 6 specimens is tested individually. After mounting the specimen on the end grips, a tensile load is applied by pulling one end at a strain rate of 3 percent per minute (1.0 millimeter per minute), while the other end remains fixed, until the specimen fails (Figure 4.63). A test typically requires less than a minute from load application until failure. A test result is considered legitimate when fracture occurs within the midsection of the specimen. A specimen that breaks near the end insert should be reported as an "end break" (Figure 4.64).

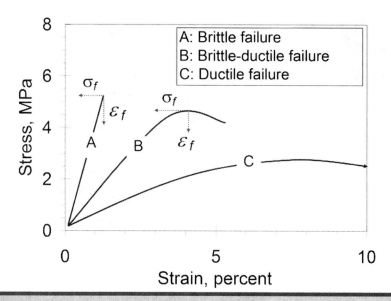

Figure 4.58 DTT stress curves

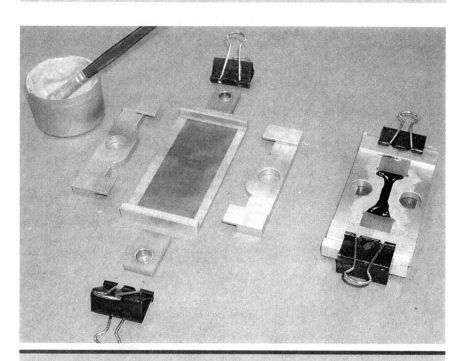

Figure 4.59 Mold assembly

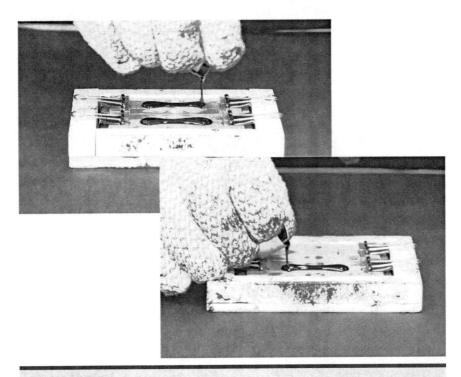

Figure 4.60 Pouring

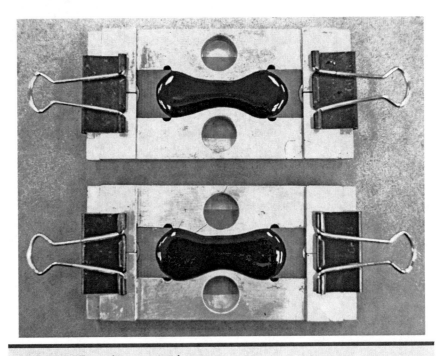

Figure 4.61 DT specimens poured

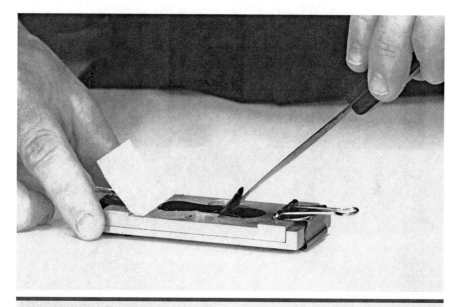

Figure 4.62 Trimming

Figure 4.63 Loading

After all 6 specimens have been tested (Figure 4.65), the 2 lowest stress-at-failure values are eliminated and the remaining 4 values are averaged to produce an average failure stress and average failure strain. This practice is based on the concept that it is much more common in direct tension tests to have an early fracture because of specimen or procedural imperfections than an exceptionally "strong" test specimen that does not fracture early. Other analysis procedures may dictate different methods for evaluating the quality of the test data and its suitability for inclusion in the general average.

Figure 4.64 Types of breaks

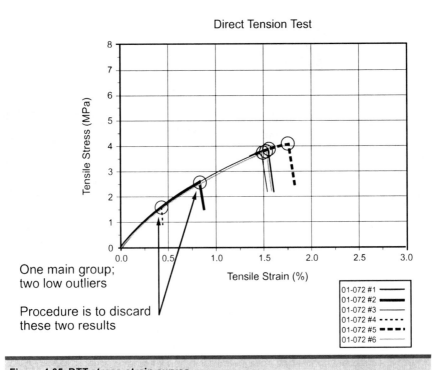

Figure 4.65 DTT stress-strain curves

A single test result consists of the average strain to failure of multiple DTT specimens, reported to the nearest 0.01 percent and the average stress at failure reported to the nearest 0.01 MPa.

Aging procedures

Two aging procedures were adopted as part of the PG Asphalt Binder Specification system to address short-term and long-term aging. Rather than developing a new procedure for short-term aging, the SHRP research team elected to use the existing Rolling Thin Film Oven Test (RTFOT) procedure. The details of the RTFOT were discussed earlier in this chapter.

The RTFO-aged asphalt binder is used for DSR testing and is transferred into small containers for additional long-term aging and future use.

As described previously, the effects of long-term, in-service aging of asphalt were not incorporated into specifications for asphalt binders before the SHRP research and the PG Asphalt Binder Specification. To address this shortcoming, the SHRP research team used an existing research procedure and equipment, the Pressure Aging Vessel (PAV), and modified it to simulate long-term, in-service aging. The PAV exposes the asphalt binder to air at high pressure and temperature for 20 hours to simulate the effects of long-term aging. Because asphalt that has been exposed to long-term aging has also been through the mixing and construction process, the PAV procedure uses asphalt binder that has been previously aged in the RTFO.

The pressure-aging apparatus consists of the pressure-aging vessel within an oven to maintain temperature (Figure 4.66). Air pressure is provided by a cylinder of clean, dry compressed air with a pressure regulator, release valve, and a slow-release bleed valve. The pressure vessel is designed to operate under the pressure (2.1 MPa) and temperature conditions (90°C, 100°C, or 110°C) of the test. A sample rack, capable of holding ten sample pans, fits conveniently into the vessel.

To prepare for the PAV, RTFO-aged binder is heated until fluid and stirred to ensure homogeneity. 50 grams of the heated asphalt binder is then poured into each of the PAV sample pans (Figure 4.67), which are then loaded in the sample rack.

The unpressurized PAV is preheated to the desired test temperature. After preheating, the sample rack with the samples is placed in the hot vessel and the lid is quickly secured to avoid excessive heat loss. When the vessel temperature is within 20°C of the required temperature, the pressure is applied and the timing for the aging period begins. After 20 hours, the pressure is gradually released, requiring 8 to 10 minutes to minimize the potential for sudden depressurization and foaming. The sample rack is then removed from the PAV. Sample pans are removed from the rack and placed in an oven at 163°C for 15 minutes before being scraped into a separate container for each asphalt binder (Figure 4.68). Note that more than 1 pan may be aged for each asphalt binder depending on the amount of aged asphalt binder needed for testing.

Figure 4.66 PAV

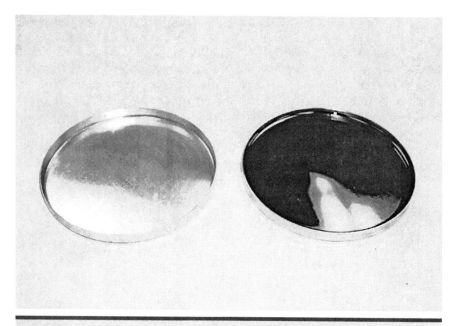

Figure 4.67 PAV pans

Figure 4.68 Scraping

MS-26

If required by the test procedure, PAV-aged residue may be subjected to a vacuum degassing procedure. It is conducted in a 170°C vacuum oven for 30 minutes at a pressure of 15 kiloPascal (absolute) to remove air bubbles from the aged asphalt binder sample. Research has shown that this is a necessary step if the asphalt binder sample will be tested using the DTT. Failure to remove the air bubbles from a PAV-aged asphalt binder could result in lower stress and strain at failure because of DTT specimen imperfections caused by air bubbles.

The PAV procedure does not produce a test result; it is merely a long-term aging procedure. Asphalt binder that has been PAV-aged is used in dynamic shear rheometer, bending beam rheometer, and direct tension tests.

PG asphalt binder specification

The central theme of the Performance Graded (PG) Asphalt Binder Specification is its reliance on testing asphalt binders in conditions that simulate critical stages during the life of the asphalt binder. Tests performed on the original asphalt binder represent the first stage of transport, storage, and handling. The second stage represents the asphalt binder during mix production and construction and is simulated for the specification by aging the asphalt binder using the rolling thin-film oven (RTFO). This procedure exposes thin binder films to heat and air and approximates the aging of the asphalt during mixing and construction. The third stage occurs as the asphalt binder ages over a long period as part of the hot mix asphalt pavement layer. This stage is simulated for the specification by aging using the pressure-aging vessel (PAV). This procedure exposes asphalt binder samples to heat and pressure to accelerate oxidation and simulate years of in-service aging in a pavement (Figure 4.69).

The PG Asphalt Binder Specification and the test methods used to characterize asphalt binders are described in AASHTO M 320. Table 1 (Figure 4.70) of the specification provides requirements for PG asphalt binders using the previously described set of tests and aging procedures. Original, unaged asphalt binder tests include the flash point, rotational viscosity, and DSR tests. RTFO aging is conducted to determine mass loss and provide short-term-aged asphalt binder for another high temperature DSR test and additional aging. PAV aging is conducted to provide long-term-aged asphalt binder for intermediate temperature DSR and low temperature BBR tests.

While the direct tension test is rarely used in Table 1, it is an integral part of Table 2 requirements. In AASHTO M 320, Table 2 requirements

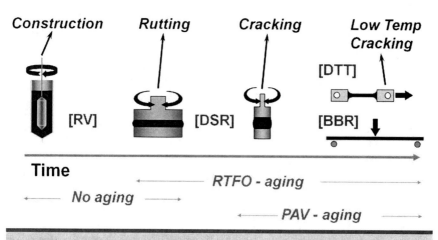

Figure 4.69 PG binder tests with aging

Table 1	Performance Grades																					
	PG 46			PG 52				PG 58				PG 64				PG 70				PG 76	PG 82
Avg. 7-day Max, °C	34	40	46	10	16	22	28	34	40	46	16	22	28	34	40	10	16	22	28	34	40	10 16 22 28 34 / 10 16 22 28 34
Original																						
≥ 230 °C	Flash Point																				
≤ 3 Pa·s @ 135 °C	Rotational Viscosity																				
≥ 1.00 kPa	DSR G*/sin δ (Dynamic Shear Rheometer)																				
	46			52					58				64				70				76	82
(Rolling Thin Film Oven) RTFO, Mass Change ≤ 1.00%																						
≥ 2.20 kPa	DSR G*/sin δ (Dynamic Shear Rheometer)																				
	46			52					58				64				70				76	82
(Pressure Aging Vessel) PAV																						
20 hours, 2.10 MPa	90			90					100				100				100 (110)				100 (110)	100 (110)
≤ 5000 kPa	DSR G*/sin δ (Dynamic Shear Rheometer)																			Intermediate Temp. = [(Max + Min)/2] + 4		
	10	7	4	25	22	19	16	13	10	7	25	22	19	16	13	31	28	25	22	19	16	34 31 28 25 22 / 37 34 31 28
S ≤ 300 MPa, m ≥ 0.300	BBR S (creep stiffness) & m-value (Bending Beam Rheometer)																				
	-24	-30	-36	0	-6	-12	-18	-24	-30	-36	-6	-12	-18	-24	-30	0	-6	-12	-18	-24	-30	0 -6 -12 -18 -24 / 0 -6 -12 -18 -24
ε ≥ 1.00%	DTT (Direct Tension Tester)																				
	-24	-30	-36	0	-6	-12	-18	-24	-30	-36	-6	-12	-18	-24	-30	0	-6	-12	-18	-24	-30	0 -6 -12 -18 -24 / 0 -6 -12 -18 -24

Figure 4.70 PG specifications table

are identical to Table 1 requirements except for low temperature testing. In Table 2, an alternate procedure is used for determining the critical low cracking temperature of an asphalt binder. It combines results from the bending beam rheometer and direct tension test procedures. Some researchers believe that the alternate approach in Table 2 is more rigorous and inclusive of most modified asphalts.

To determine or verify the low temperature grade of an asphalt binder using the procedures in Table 2, the technologist needs to have BBR data at a minimum of two test temperatures. Although it is not required, it is preferable that one of the test temperatures produces passing BBR data—$S(60) \leq 300$ MPa and $m(60) \geq 0.300$—and the other temperature produces failing BBR data. The BBR data from each of the loading times (not just 60 seconds) is input into a spreadsheet or commercially available software to generate a thermal stress curve. This thermal stress curve describes how stress is expected to increase in an asphalt binder as the temperature decreases.

In addition to the BBR data (used to generate the thermal stress curve), the technologist needs to have failure stress data from the direct tension test at one or more temperatures. If just one temperature is available, the technologist can validate that the asphalt binder either does or does not meet the required low temperature grade. If two or more temperatures are available, the technologist can determine the continuous low temperature grade by determining the critical cracking temperature (CCT) as shown in Figure 4.71.

In Figure 4.71, the intersection of the failure stress curve from the DTT and the thermal stress curve from the BBR occurs at a temperature defined as the critical cracking temperature, CCT. If DTT data is only available at one temperature, like $-12°C$, then the technologist can only determine if the asphalt binder meets that low temperature grade by examining if the failure stress occurs above or below the thermal stress curve. If the point is above the curve, as is the data point at $-12°C$ in Figure 4.71, then the

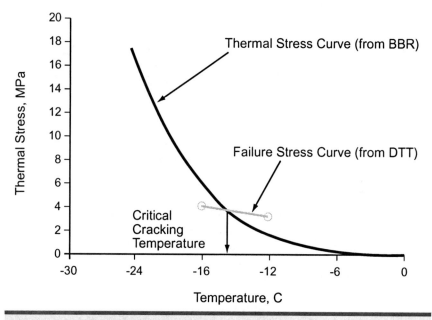

Figure 4.71 Critical cracking temperature

thermal stress does not exceed the failure stress and the asphalt binder passes the grade criterion at that temperature. If the point is below the curve, as is the data point at −18°C in Figure 4.71, then the thermal stress exceeds the failure stress and the asphalt binder fails the grade criterion at that temperature.

Advantages compared with older specifications

A unique feature of the PG Asphalt Binder Specification is that, for the DSR and BBR tests, the specified criteria remain constant, but the temperature at which the criteria must be achieved changes for the various grades. As an example, consider two construction projects—one at the equator and one at the Arctic Circle. Good performance is expected in both locations, but the temperature conditions under which specified asphalt binder properties must be achieved are vastly different. This approach differs from previous specification systems, which relied on measuring a property at a fixed temperature but allowed the criteria to change based on the climate in which the asphalt binder would be used.

The PG asphalt binder tests measure physical properties that can be related directly to field performance by engineering principles. The use of complex shear modulus at high test temperatures allows for a better characterization of both the elastic and viscous components of the asphalt binder. This makes the test more appropriate than the absolute

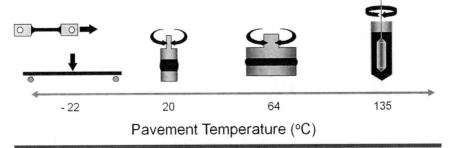

Figure 4.72 PG binder tests with temperature

viscosity test for modified asphalt binders, considering that many have a substantial elastic component at high temperature.

Unlike previous specifications, the PG asphalt binder tests are conducted at the entire temperature range expected to be encountered by the asphalt binder during production and in-service in the asphalt pavement (Figure 4.72). Previous systems only routinely characterized physical properties as low as 77°F (25°C), with a few tests conducted at 41°F (4°C). Because asphalt pavements are frequently exposed to much lower temperatures, the ability to measure physical properties at a low temperature using the BBR and direct tension tests is a significant advancement.

Finally, the adoption of the pressure-aging vessel procedure to simulate the long-term aging of asphalt binders is also an advancement in the characterization of asphalt binder properties. Fatigue and thermal cracking are most likely to occur in an asphalt pavement after the asphalt binder has been in service for several years. The pressure-aging vessel procedure simulates this in-service aging, allowing for testing to be conducted at intermediate and low temperatures to determine the physical properties of the asphalt binder believed to be related to these distresses.

Assumptions in the specification

The PG Asphalt Binder Specification incorporates several simplifying assumptions that permit the use of properties such as complex shear modulus (G^*). Asphalt binders that violate these assumptions cannot be effectively graded in the PG Asphalt Binder Specification. The asphalt binders must be rheologically-simple systems. What does this mean?

First, the asphalt binders should be homogeneous. Particulate matter that has a particle size greater than 0.25 millimeters may interfere with the consistent response of tests, such as the DSR test at a high temperature. The size of such particles may cause the G^* measurement to be inaccurate and display behavior that is not linear viscoelastic. Asphalt binders with particulate additives may not meet this assumption.

Although not listed in Table 1 or Table 2, the asphalt binder must have a solubility of 99 percent or more according to AASHTO M 320. This requirement helps to ensure that excess inorganic material is not present in the asphalt binder.

Second, the asphalt binders should exhibit isotropic behavior. Isotropic behavior occurs when specimen loading or particle orientation has no effect on the response. Asphalt binders that incorporate fibers could exhibit anisotropic behavior—meaning the fiber orientation affects the test response.

Third, PG asphalt binders should exhibit linear viscoelastic behavior over a wide range of strain. Linear viscoelastic behavior can be determined by performing a strain sweep with the DSR. In this procedure, the asphalt binder sample is tested at several increasing strain levels with G^* determined at each strain level. To be a linear viscoelastic material, G^* should remain relatively constant over a wide range of shear strain.

How the PG specification addresses pavement performance

Safety

In the PG specification, safety is accounted for by requiring a minimum flash point temperature using the Cleveland Open Cup Flash Point test. This test procedure (AASHTO T 48) indicates the temperature at which an asphalt binder will instantaneously flash in the presence of an open flame. The flash point is well below the fire point, or the temperature at which the asphalt binder will actually burn.

The minimum temperature required for all PG asphalt binders is 230°C (approximately 450°F). This test is performed on the unaged binder.

Pumping and handling

To ensure that asphalt binders, especially modified asphalts, can be pumped and handled at the hot-mixing facility, the PG specification contains a maximum viscosity requirement on the unaged binder as determined by AASHTO T 316. This value is 3.0 Pascal-seconds (Pa-s) and it must be achieved at 135°C for all grades. The specifying agency may waive this requirement if the supplier warrants that the binder can be handled and pumped at the necessary temperatures.

While the limiting viscosity of 3.0 Pa-s is not a problem for most PG asphalt binders, it may be exceeded for some highly modified asphalt binders. Users requesting an asphalt binder having a useful temperature interval, UTI, (temperature range from high-temperature grade to low-temperature grade) of greater than 100 degrees should expect the increased likelihood that the supplied asphalt binder will exceed the rotational viscosity limit of 3.0 Pa-s (3000 centipoise) at 135°C. Exceeding the limit in these instances should not be considered a material property failure since this is simply a result of the high modification level needed to achieve the required temperature spread.

The user agency, hot mix asphalt contractor, and asphalt binder supplier should work together to ensure that the required grade is appropriate for the project and that the supplied asphalt binder can be properly handled by the hot mix asphalt contractor to produce a mixture having the desired in-place properties.

Permanent deformation

As discussed in the section describing the DSR, the total response of an asphalt binder to load consists of elastic (recoverable) and viscous (non-recoverable) components. Pavement rutting, or permanent deformation, is the accumulation of nonrecoverable deformation in the asphalt mixture in response to repeated load applications at high temperatures (Figure 4.73). To address the response of the asphalt binder to these repeated loads, the PG specification defines and places requirements on a rutting factor, $G^*/\sin \delta$ (read as "G star over sine delta."), which represents a measure of the high-temperature stiffness of the asphalt binder. This factor is determined by dividing the complex shear modulus (G^*) by the sine of the phase angle (δ), as determined using the dynamic shear rheometer (DSR) test procedure (AASHTO T 315). To minimize the contribution of the asphalt binder to rutting, the value of $G^*/\sin d$ must be a minimum of 1.00 kiloPascal for the original asphalt binder and 2.20 kiloPascal for the short-term-aged asphalt binder (after using the RTFO procedure) at the appropriate grade temperature.

A minimum RTFO $G^*/\sin d$ value ensures that the asphalt binder will have sufficient stiffness after construction to minimize the contribution of the asphalt binder to permanent deformation in the asphalt mixture. High values of G^*

Figure 4.73 Rutting

and low values of δ are considered desirable attributes from the standpoint of rutting resistance. Thus, the PG specification promotes the use of stiff, elastic binders (unaged and RTFO-aged) to address permanent deformation.

Excessive aging

As with previous grading systems, a mass-loss requirement is specified to help prevent the use of an asphalt binder that would age excessively from volatilization during hot mixing and construction. The mass-loss requirement is calculated using the RTFO procedure (AASHTO T 240). The mass loss for any PG asphalt binder should not exceed 1.00 percent.

Long-term stiffness

Like permanent deformation, G^* and δ are also used in the PG specification to characterize asphalt binder stiffness after long-term aging, when subjected to cumulative loading at moderate pavement temperatures. Excessive stiffness at intermediate temperatures could be a contributing factor in cracking (durability) of an asphalt pavement (Figure 4.74). Because cracking generally occurs at lower to moderate pavement temperatures after the pavement has been in service for some time, the specification addresses the response of the asphalt binder after aging by both the RTFO and PAV.

The DSR is again used to generate G^* and δ. However, instead of dividing the two parameters, the two are multiplied to produce a factor that may be related to the long term oxidation and aging properties of an asphalt binder at intermediate temperatures. The factor, *$G^*\sin δ$*, (read as "G star sine delta"), is the product of the complex shear modulus, G^*, and the sine of the phase angle, δ. The PG specification requires that all PG asphalt binders have a maximum value of 5000 kiloPascal for $G^*\sin δ$ at the appropriate grade temperature.

The maximum value of 5000 kiloPascal was considered to be an acceptable specification limit when the PG specification was first adopted, based on the observed fatigue cracking performance of asphalt pavement sections used in the Zaca-Wigmore test sections. Low values of $G^*\sin δ$ indicate low energy dissipation. Thus, a maximum allowable value was considered for the specification. Subsequent research has not validated that the $G^*\sin δ$ parameter is related to fatigue performance. Rather, the parameter is considered to provide an indication of stiffness at intermediate temperatures as a result of long-term aging.

Thermal cracking

When temperature decreases, asphalt pavements contract, causing thermal stress to develop in the pavement. As the temperature drops, the asphalt binder contracts to a much greater degree than the aggregate in an asphalt pavement. When these stresses exceed the tensile strength of the asphalt mixture, a low-temperature crack develops (Figures 4.75 and 4.76). Asphalt becomes more brittle with time, which is why thermal cracking occurs at low pavement temperatures after the pavement has been in service for a time. Because of this, the specification addresses the response of the asphalt binder after aging in both the RTFO and PAV.

Figure 4.74 Fatigue cracking

Figure 4.75 Thermal cracking—early

Figure 4.76 Thermal cracking—late

In the PG specification, the main way of examining the propensity of an asphalt binder to develop thermal stresses at a specified low temperature is to use data generated from the BBR. If the stiffness is too high, the asphalt binder will behave in a brittle manner, indicating that cracking is more likely to occur. To minimize the contribution of the asphalt binder to thermal cracking, the creep stiffness *(S)* after 60 seconds of loading at the appropriate temperature must not exceed 300 MPa. In addition to stiffness, the rate at which the asphalt binder stiffness changes with time at low temperatures is regulated through the *m-value*. In the PG specification, a higher *m-value* is an indication that the asphalt binder may not increase in stiffness as rapidly when the temperature decreases and contraction occurs, leading to smaller tensile stresses in the asphalt binder and less chance for low temperature cracking. A minimum *m-value* of 0.300 after 60 seconds of loading at the appropriate temperature is required by the PG specification.

Past studies have also indicated that if a binder can stretch at least 1.0 percent of its original length during this thermal contraction period, cracks are less likely to occur. As a result, the direct tension test (DTT) is included in the PG specification as an alternative requirement to using creep stiffness and *m-value*. In the specification, the direct tension test requirement only applies to asphalt binders that have a creep stiffness greater than 300 MPa but less than 600 MPa, with an *m-value* of 0.300 or greater. If the creep stiffness at the specified temperature is 300 MPa or less, then the direct tension test is not required. If the direct tension test is used, the failure strain must be a minimum of 1.0 percent at the appropriate temperature.

Although stiffness can also be used to estimate failure or strength properties, for some asphalt binders, especially modified asphalts, the relationship between stiffness and strength properties is not well known. This is why some researchers believe that the alternate procedure for determining a critical low cracking temperature specified in Table 2 of the AASHTO M 320 specification represents a more rigorous approach than the use of BBR stiffness and *m-value* alone (as is commonly used in the PG specification).

Grading an asphalt binder in the PG specification system

Unlike previous grading systems, the PG Asphalt Binder specification is based directly on performance-related properties rather than on empirical relationships between basic physical properties and observed performance. Performance graded (PG) asphalt binders are selected based on the climate in which the pavement will serve. Unlike other systems, the physical property requirements (representing good performance) are constant among all grades. The distinction among the various binder grades is the specified temperatures at which the requirements must be met. For example, a binder classified as a PG 58-34 means that the binder will meet the high temperature physical property requirements up to a temperature of 58°C and the low temperature physical property requirements down to a temperature of −34°C.

AASHTO M 320 contains a listing of the more common PG binders. However, the PG binders are not limited to those given classifications. In actuality, the specification temperatures are unlimited, extending unbounded in both directions. The high and low temperatures extend as far as necessary in the standard 6-degree increments. For example, even though a PG 58-10 is not shown, it exists as a legitimate grade in the system. Similarly, some states in the southeastern U.S. have elected to use a PG 67-22 asphalt binder. Even though it is not listed as a standard grade in AASHTO M 320, the same grading concepts still apply.

The first number listed in a PG asphalt binder grade is the high temperature grade. It represents the maximum pavement design temperature where the asphalt binder may be expected to perform. The second number listed in a PG asphalt binder grade is the low temperature grade. It represents the minimum pavement design temperature where the asphalt binder may be expected to perform. Thus a PG 58-22 asphalt binder would be expected to provide a reasonable assurance of performance in a climate where the high pavement temperature does not exceed 58°C and the low pavement temperature does not drop below −22°C. Note that the low temperature grade is most often a negative number, so the dash ("-") is actually a minus sign.

Non-grade determination requirements

Separate from the temperature grading are certain requirements that must be met for all asphalt binders in AASHTO M 320. If these requirements are not met, the asphalt binder cannot be considered to meet the AASHTO M 320 requirements.

Although it is not listed in either Table 1 or Table 2, the text in the specification indicates that the asphalt binder must have a solubility of 99.0 percent or greater. Although AASHTO T 44 is specified, users should be aware that some modified asphalt binders may not meet the solubility requirement because of the type of modifier and solvent used.

As noted previously, all asphalt binders must exhibit a mass change of 1.00 percent or less after RTFO aging. This requirement ensures that excessive volatilization and aging doesn't occur in the short-term.

Asphalt binders must have a rotational viscosity of 3 Pa-s or less when measured at 135°C. This is primarily a requirement to ensure that the asphalt binder is not so stiff that it cannot be effectively pumped and handled.

Finally, as a safety precaution, asphalt binders are required to have a flash point (measured with the Cleveland Open Cup) of at least 230°C.

Verifying asphalt binder Performance Grade using AASHTO M 320 Table 1

AASHTO R 29 provides guidelines for grading or verifying the grade of an asphalt binder. ***Grade verification*** is a process where the asphalt binder is tested to determine conformance with a specified grade. In this case, a sample is supplied to the testing lab and the grade is identified by the supplier. The testing lab then tests the sample at the appropriate test temperatures and determines if the asphalt binder conforms to the expected grade.

In addition to the non-grade determination requirements, the asphalt binder must meet the grade determination requirements. For a PG 70-22 asphalt binder this means that the asphalt binder must have:
- a G*/sin δ value on the original, unaged asphalt binder that meets or exceeds 1.00 kiloPascal at 70°C; and
- a G*/sin δ value on the RTFO-aged asphalt binder that meets or exceeds 2.20 kiloPascal at 70°C; and
- a G*sin δ value on the PAV-aged asphalt binder that is no greater than 5000 kiloPascal at 28°C (required intermediate temperature from AASHTO M 320 Table 1 for a PG 70-22 grade); and
- a BBR Stiffness, S(60), on the PAV-aged asphalt binder that is no greater than 300 kiloPascal at −12°C; and
- a BBR m-value, m(60), on the PAV-aged asphalt binder that is at least 0.300 at −12°C.

Note that the low temperature tests are conducted at a temperature that is 10°C warmer than the actual low temperature grade. As such a PG xx-28 asphalt binder is tested at −18°C (instead of −28°C) and a PG xx-16 asphalt binder is tested at −6°C (instead of −16°C)

Grading an unknown sample

If the grade of the asphalt binder sample is not known, or if the actual passing and failing temperatures are desired, then AASHTO R 29 provides for **Grade classification** of an asphalt binder. **Grade classification** is a process where the asphalt binder is tested to determine the temperatures at which the asphalt binder meets the physical property requirements. It is similar to **Grade verification**, but, in this case, a sample is supplied to the testing lab and the grade is not identified by the supplier. The testing lab then systematically tests the sample by starting at a temperature and adjusting test temperatures up or down depending on the test results. If the sample meets the specification criteria for high temperature tests, then the test temperature is *increased 6 degrees* and the sample is tested again. If the sample meets the specification criteria for low temperature tests, then the test temperature is *decreased 6 degrees* and the sample is tested again. If the sample meets the specification criteria for intermediate temperature tests, then the test temperature is *decreased 3 degrees* and the sample is tested again. Experienced technicians can usually pick an appropriate starting temperature based on the consistency and appearance of the asphalt material. However, in the absence of experience, a good starting temperature is 58°C for high temperature tests, 25°C for intermediate temperature tests, and −12°C for low temperature tests.

Approximations/rules of thumb

As a general rule of thumb, asphalt binder stiffness will change by a factor of 2 for every 6°C change in temperature. If the temperature increases by

6°C, then the stiffness will be reduced by half; if the temperature decreases by 6°C, then the stiffness will be doubled.

This is a very useful rule of thumb since it allows a quick estimation of the temperature at which the asphalt binder will meet the specification criteria. As an example, suppose an asphalt binder sample is tested and has an original G*/sin δ value of 1.62 kiloPascal at 64°C. The technician knows that the asphalt binder will almost assuredly not meet the specification criterion of 1.00 kiloPascal at 70°C, since the stiffness will be reduced by half. Therefore, unless a failing temperature is needed, the technician can reasonably assume that the original high temperature grade will be a PG 64.

The *m-value* is somewhat different in its response to temperature changes since it is not stiffness but a rate of stiffness change. Generally *m-value* can be expected to change by 0.050 for every change of 6°C in temperature. If an asphalt binder has an *m-value* of 0.322 at −12°C, the technician should expect to see an *m-value* of approximately 0.272 at −18°C. Caution should be used in applying this rule of thumb for *m-value* as it is an approximation based on conventional PAV-aged asphalt binders. Modified asphalt binders and asphalt binders that have been significantly oxidized (such as after long-term, in-service aging) may not adhere to this general approximation.

Critical temperature, continuous grade, and UTI

In asphalt binder grading, it may be necessary to determine the actual temperatures at which the specification criteria are met. Typically this is done to allow for linear blending equations to be used. During the **Grade classification** process, results are usually obtained at: (1) a temperature where the specification criteria are met; and (2) the next higher or lower temperature where the specification criteria are not met. By obtaining passing and failing temperatures, the temperature where the specification is exactly met can be determined. This is referred to as the **critical temperature (T_c)**. While the **critical temperature** can be determined using two passing or two failing temperatures, the calculation then would involve extrapolation. Whenever possible, it is better to interpolate between two data points than to extrapolate outside of the data. The equations for determining **critical temperature** are shown below:

Original DSR

$$T_c = T_1 + \left[\frac{Log(1.00) - Log(G_1^*/\sin \delta_1)}{Log(G_1^*/\sin \delta_1) - Log(G_2^*/\sin \delta_2)} \times (T_1 - T_2)\right]$$

In the preceding equation, T_1 and T_2 are the two measurement temperatures and $G_1^*/\sin \delta_1$ and $G_2^*/\sin \delta_2$ are the original DSR measurements at T_1 and T_2, respectively.

RTFO DSR

$$T_c = T_1 + \left[\frac{Log(2.20) - Log(G_1^*/\sin \delta_1)}{Log(G_1^*/\sin \delta_1) - Log(G_2^*/\sin \delta_2)} \times (T_1 - T_2)\right]$$

In the preceding equation, T_1 and T_2 are the two measurement temperatures and $G_1^*/\sin \delta_1$ and $G_2^*/\sin \delta_2$ are the RTFO DSR measurements at T_1

and T_2, respectively. Note that the equation is the same as the T_c equation for Original DSR except that the criterion changes (2.20 instead of 1.00 kiloPascal).

PAV DSR

$$T_c = T_1 + \left[\frac{Log(5000) - Log(G_1^* \sin \delta_1)}{Log(G_1^* \sin \delta_1) - Log(G_2^* \sin \delta_2)} \times (T_1 - T_2)\right]$$

In the preceding equation, T_1 and T_2 are the two measurement temperatures and $G_1^* \sin \delta_1$ and $G_2^* \sin \delta_2$ are the PAV DSR measurements at T_1 and T_2, respectively.

BBR Stiffness

$$T_c = T_1 + \left[\frac{Log(300) - Log(S_1)}{Log(S_1) - Log(S_2)} \times (T_1 - T_2)\right] - 10$$

In the preceding equation, T_1 and T_2 are the two measurement temperatures and S_1 and S_2 are the BBR Stiffness measurements at T_1 and T_2, respectively. Note that the form of the equation is essentially the same as the preceding equations except that 10 is subtracted from the calculations to account for the fact that low temperature testing in AASHTO M 320 is conducted at a temperature that is 10°C higher than the grade temperature of the asphalt binder.

BBR m-value

$$T_c = T_1 + \left[\frac{0.300 - m_1}{m_1 - m_1} \times (T_1 - T_2)\right] - 10$$

In the preceding equation, T_1 and T_2 are the two measurement temperatures and m_1 and m_2 are the BBR *m-value* measurements at T_1 and T_2, respectively. Note that the form of the equation is different from the preceding equations because *m-value* is a rate of change of log stiffness versus log time. As such, the equation is a linear interpolation between two data points. Note also that 10 is subtracted from the calculations to account for the fact that low temperature testing in AASHTO M 320 is conducted at a temperature that is 10°C higher than the grade temperature of the asphalt binder.

By using the **critical temperature** for each test condition, we can determine the **continuous grade** of the asphalt binder. This can be very useful for blending or for simply understanding the true properties of the asphalt binder. As an example, consider the data in Table 4.2. Both asphalt binders have an AASHTO M 320 grade of PG 64-22, but the expected performance of the two could be somewhat different since the **continuous grades** are different.

When there are more than one set of specification requirements—such as Original and RTFO G*/sin δ to determine the high temperature grade—then the most restrictive of the two parameters is selected. In Table 4.2, for both asphalt binders, the m(60) criterion is met at a warmer temperature than the S(60) criterion. Therefore, the m(60) temperature is used to determine the low temperature continuous grade.

Table 4.2
Comparison of AASHTO M 320 grade with continuous grade

Property	Binder A	Binder B
Original G*/sin δ		
@ 64°C	1.12 kPa	1.75 kPa
@ 70°C	0.55 kPa	0.90 kPa
RTFO G*/sin δ		
@ 64°C	2.53 kPa	3.65 kPa
@ 70°C	1.43 kPa	1.84 kPa
PAV G*sin δ		
@ 25°C	3120 kPa	2850 kPa
@ 22°C	3980 kPa	3650 kPa
PAV BBR S(60)		
@ −12°C	188 MPa	156 MPa
@ −18°C	375 MPa	320 MPa
PAV BBR m(60)		
@ −12°C	0.318	0.339
@ −18°C	0.270	0.293
AASHTO M 320 Grade	PG 64-22	PG 64-22
Critical Temp., T_c, °C		
Original G*/sin δ	65	69
RTFO G*/sin δ	65	68
High Temperature	65	68
Int. Temperature	19	18
BBR S(60)	−26	−27
BBR m(60)	−24	−27
Low Temperature	−24	−27
Continuous Grade	PG 65-24	PG 68-27
UTI	89°C	95°C

The last row in Table 4.2 is UTI, or Useful Temperature Interval. This value describes the range of temperature at which the asphalt binder meets specification criteria. The greater the UTI, the greater the temperature range in which the asphalt binder meets specification criteria.

Asphalt binder grade selection

Even with binder grades classified according to high and low temperature categories, more information is needed to select an asphalt binder for a particular location. The LTPPBind software, developed by the Federal Highway Administration, assists users in selecting asphalt binder grades for their pavement/climate conditions. It is important to note that it is not necessary to use the LTPPBind software to select an asphalt binder grade. Many agencies have already evaluated their climatic conditions and expected traffic loading and have selected appropriate grades. The LTPPBind software is simply a tool that can help users better understand the air/pavement temperatures that are expected for their given project location.

The LTPPBind software provides a database of weather information for nearly 6,500 reporting weather stations in the U.S. and Canada that allows users to select binder grades for the climate at the project location. For each year that these weather stations have been in operation, the hottest 7-day period was identified and the average maximum air temperature for this 7-day period was calculated. SHRP researchers selected this 7-day average value as the optimum method to characterize the high temperature design condition. For all the years of operation, the mean and standard deviation of the 7-day average maximum air temperature have been computed. Similarly, the 1-day minimum air temperature of each year was identified and the mean and standard deviation of all the years of record were calculated. Weather stations with less than 20 years of operations were not used.

Why use a 7-day maximum for high temperature and a 1-day minimum for low temperature? The researchers reasoned that high temperature damage (rutting) was caused by a combination of repeated loads at high temperatures. Thus, it was necessary to account for the accumulated damage caused by the combination of repeated loads. Low temperature damage (cracking) can be the result of a single thermal event and is not dependent on traffic loading. Thus, a single cold temperature could initiate a crack.

Because weather stations record air temperatures, it is necessary to convert to pavement temperatures before selecting an asphalt binder grade. For surface layers, researchers defined the high pavement design temperature at a depth 20 millimeters below the pavement surface and the low pavement design temperature at the pavement surface.

Why calculate the high pavement temperature at a depth of 20 millimeters and the low pavement temperature at the pavement surface? In this case, it is a matter of choosing the most conservative condition for determining damage. For rutting, research determined that the temperature decreased continually as the depth from the surface increased, but shear stresses created by traffic increased to a maximum value at a depth of approximately 50 millimeters before decreasing further with depth. The 20-millimeter depth represents the location where the combined effects of temperature and shear stress were greatest. For low temperature cracking, which is only affected by temperature, the temperature at the pavement surface is the lowest, and therefore the critical condition.

Using theoretical analyses of actual conditions performed with models for net heat flow and energy balance, and assuming typical values for solar absorption (0.90), radiation transmission through air (0.81), atmospheric radiation (0.70), and wind speed (4.5 m/sec), an equation was developed for the high pavement design temperature:

$$T_{20mm} = (T_{air} - 0.00618*Lat^2 + 0.2289*Lat + 42.2)*(0.9545) - 17.78$$

where
    T_{20mm} = high pavement design temperature at a depth of 20 millimeters
    T_{air} = 7-day average high air temperature, °C
    Lat = the geographical latitude of the project in degrees.

The low pavement design temperature at the pavement surface is calculated using the equation:

$$T_{pav} = -1.56 + 0.72*T_{air} - 0.004*Lat^2 + 6.26 \log_{10}*(H + 25) - Z*(4.4 + 0.52\, \sigma^2_{air})^{1/2}$$

where
- T_{pav} = Low AC pavement temperature below surface, °C
- T_{air} = Low air temperature, °C
- Lat = Latitude of the section, degrees
- H = Depth to surface, millimeter
- σ_{air} = Standard deviation of the mean low air temperature, °C
- Z = from the standard normal distribution table, Z = 2.055 for 98 percent reliability.

The system allows the designers to use reliability measurements to assign a degree of design risk to the high and low pavement temperatures used in selecting the asphalt binder grade. As defined in the Superpave system, reliability is the percent probability that in a single year the actual temperature (1-day low or 7-day average high) will not exceed the design temperatures. A higher reliability means lower risk. For example, consider summer air temperatures in Cleveland, Ohio, which has a mean 7-day maximum of 32°C and a standard deviation of 2°C. In an average year, there is a 50 percent chance that the 7-day maximum air temperature will exceed 32°C. However, assuming a normal statistical frequency distribution, there is only a 2 percent chance that the 7-day maximum will exceed 36°C (mean plus two standard deviations); therefore, as shown in Figure 4.77, a design air temperature of 36°C will provide 98 percent reliability.

Continuing the example, assume that an asphalt mixture is to be designed for a project in Cleveland. Figure 4.78 graphically represents the statistical variation of the two design air temperatures. In a normal

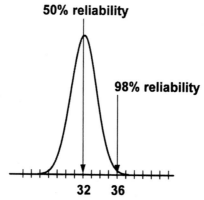

Figure 4.77 Design temperature reliability

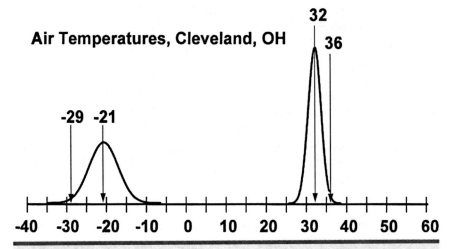

Figure 4.78 High and low air temperature variations

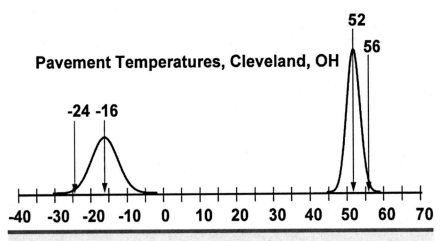

Figure 4.79 High and low pavement temperature variations

summer, the average 7-day maximum air temperature is 32°C, and in a very hot summer this average may reach 36°C. Using a similar approach for winter conditions, Cleveland has an average minimum air temperature of −21°C with a standard deviation of 4°C. Consequently, in an average winter, the coldest temperature is −21°C. For a very cold winter, the air temperature may reach −29°C. The standard deviations show there is more variation in the 1-day low temperatures than the 7-day average high temperatures.

For Cleveland (latitude = 41.42 degrees) the design pavement temperatures are calculated to be approximately 52°C and −16°C for 50 percent reliability and approximately 56°C and −23°C for 98 percent reliability (mean plus two standard deviations). Figure 4.79 graphically represents the statistical variation of the two pavement temperatures.

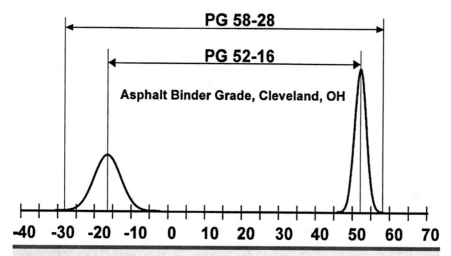

Figure 4.80 PG asphalt binder grade selections

To achieve a reliability of at least 50 percent and provide for an average maximum pavement temperature of at least 52°C, the high temperature grade happens to match the design temperature, PG 52. Using the same reasoning, the low temperature grade is a PG −16 to attain 50 percent reliability. Coincidentally, the low temperature grade again happens to match the design temperature, −16. As shown in Figure 4.80, to obtain at least 98 percent reliability, it is necessary to select a high temperature grade of PG 58 to protect above 56°C and a low temperature grade of PG −28 to protect below −23°C. In both the high and low temperature cases of the PG 58-28 binder grade, the actual reliability exceeds 99 percent because of the rounding up caused by the six-degree difference between standard grades in AASHTO M 320.

By rounding up to a standard grade, the user introduces conservatism into the asphalt binder selection process. Another possible source of conservatism occurs when considering the same steps encountered during asphalt binder test classification. Although a specific asphalt binder may pass all of the criteria when tested at lower or higher temperatures, it will nevertheless be classified by rounding down to the next 6-degree step of the grading system. The net result is that a significant factor of safety is included in the process of selecting an asphalt binder that is appropriate for the project conditions. For example, it is possible that the PG 52-16 binder, selected previously for a minimum of 50 percent reliability for Cleveland may have a continuous grade of PG 56-20. Users of this temperature-based stepped grading system for binder selection should recognize that considerable safeguards are already included in the process. Because of these factors, it may not be necessary or cost-effective to require indiscriminately high values of reliability or abnormally conservative high or low temperature grades.

The LTPPBind software converts air to pavement temperature with minimal user input. For any site, the user can enter a minimum reliability and the software will calculate the required asphalt binder grade. Alternatively, the user can specify a desired asphalt binder grade and the software will calculate the reliability obtained for the climate at the site. Consequently, users are faced with engineering management decisions. They will have to decide the level of reliability to be used. Depending on the policy established by each individual agency, the selected reliability may be a function of road classification, traffic level, cost, and other factors.

It should be emphasized that proper or conservative selection of an asphalt binder grade does not guarantee total pavement performance. Fatigue cracking performance is greatly affected by the pavement structure and traffic. Permanent deformation or rutting is directly a function of the shear strength of the mix, which is greatly influenced by aggregate and mixture properties, such as level of compaction. Of the major distresses, pavement low temperature cracking correlates most significantly to the binder properties. Engineers should try to achieve a balance among the many factors when selecting binders.

Grade bumping for greater traffic loading

The asphalt binder selection procedure described previously is the basic procedure for typical highway loading conditions. Under these conditions, it is assumed that the pavement is subjected to a design number of fast, transient loads. For the high temperature design situation, controlled by specified properties relating to permanent deformation, the speed of loading has an additional effect on performance. The system allows for an additional shift in the selected high temperature binder grade to account for slow transient and standing load applications. Similar to the time-temperature shift described with the test temperature for the BBR (testing at 10°C higher temperature reduced the test duration from 2 hours to 60 seconds), higher maximum temperature grades are used to offset the effect of slower loading speed. For slow moving design loads, the asphalt binder grade should be selected one high temperature grade to the right (i.e., one grade "warmer").

In the Cleveland example above, if the user selected a 98 percent minimum reliability for standard loading, a PG 58-28 asphalt binder would be required. Under conditions of slow loading, the user would shift (or bump) the high temperature grade to a PG 64, while leaving the low temperature grade unaffected. Thus a PG 64-28 would be selected instead of a PG 58-28. For standing design loads, the asphalt binder should be selected two high temperature grades to the right or two grades "warmer." Continuing with the example, a PG 70-28 would be selected instead of a PG 58-28.

Also, a shift is recommended for high numbers of heavy traffic loads. These are locations where the traffic is expected to exceed 10,000,000 equivalent single axle loads (ESAL) over the design life of the pavement. An ESAL is defined as one 18,000-pound (80-kN) 4-tired dual axle and is the unit used by most pavement thickness design procedures to quantify the various types of axle loadings into a single design traffic number. If the design traffic is expected to be between 10,000,000 and 30,000,000 ESAL, then the engineer may consider selecting one high temperature binder grade higher than the selection based on climate. If the design traffic is expected to exceed 30,000,000 ESAL, then the binder should be selected one high temperature grade higher.

Performance-Graded asphalt binder using the Multiple Stress Creep Recovery (MSCR) test

As discussed previously, one of the objectives in the development of the PG Asphalt Binder Specification was the use of performance-related criteria specific for a distress and related to climate and traffic loading. This objective implies that test measurements should be made at temperatures and loading rates consistent with conditions existing in the pavement. With this approach, the high temperature criteria stays the same for $G^*/\sin \delta$ (1.00 kiloPascal for unaged and 2.20 kiloPascal for RTFO-aged binder) regardless of the location of the pavement. However, the test temperature where this criteria must be met is derived from the actual pavement temperature.

While this concept worked well for conventional speed and moderate traffic volume pavements, research indicated that it needed some refinement for pavements that had slow speed loading and high traffic volume. Rather than change criteria and/or test conditions to reflect a change in loading time and traffic volume, the architects of the PG system elected to simply adjust for traffic speed and volume by grade-bumping or testing at higher temperatures than indicated by the climate. So for a standard traffic asphalt pavement, the designer might use a PG 58-28 asphalt binder; but a high volume highway pavement might require a PG 70-28 asphalt binder, even though the pavement temperature would likely never get above 58°C. This was a simple way to ensure that a stiffer asphalt binder would be used in high volume and/or slow loading conditions. One problem with grade-bumping in the PG system was that a PG 70-28 asphalt binder would have its performance-related properties determined at a temperature that would be at least 12°C hotter than the highest pavement temperature that would be experienced. Such high specified testing temperatures in some instances have caused asphalt suppliers to manufacture binders that are very highly modified and thus difficult to use at reasonable temperatures.

Another objective of the PG system was that the performance-related properties that defined the performance grade of an asphalt binder would be blind to modification. In other words, all asphalt binders of the same performance grade would be expected to perform the same in the same traffic/environmental conditions regardless of how they were produced. This would eliminate the specification proliferation that had become increasingly more common as modification of asphalt binders became more common. While the $G^*/\sin \delta$ parameter did capture viscous and elastic effects, it was unable to adequately capture the benefits of elastomeric modification because of the relatively small impact of phase angle (δ) on the overall value of $G^*/\sin \delta$. As a result, additional empirical tests

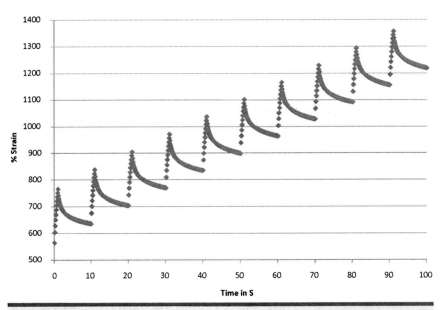

Figure 4.81 Creep-recovery response from MSCR

(sometimes referred to as "plus" tests and discussed later) continued so that a user agency could have assurance that they were getting a polymer-modified asphalt binder as they had in the past.

These issues caused researchers to continue to look for an improvement to the high temperature parameter, $G^*/\sin \delta$, used in AASHTO M 320. Their efforts led to the development of a new test procedure, the Multiple-Stress Creep-Recovery (MSCR) test.

MSCR test and specification

The MSCR test was developed based on creep and recovery work conducted on asphalt binders and mixtures. During SHRP, researchers at the University of California at Berkeley developed the Repeated Simple Shear Test at Constant Height (RSST-CH) for asphalt mixtures. The RSST-CH was developed to characterize the rutting performance of asphalt mixtures and was conducted using repeated cycles of 0.1-second shear load followed by a 0.6-second rest period. Based on this mixture work, the NCHRP 9–10 research used a repeated creep-recovery test to characterize the expected rutting performance of modified asphalt binders.

The MSCR test (AASHTO TP 70) uses the well-established creep and recovery test concept to evaluate the asphalt binder's potential for permanent deformation. Using the DSR, a 1-second creep load is applied to the RTFO-aged asphalt binder sample. After the 1-second load is removed, the sample is allowed to recover for 9 seconds. The test is started with the application of a low stress (0.1 kiloPascal) for 10 creep/recovery cycles and then the stress is increased to 3.2 kiloPascal and repeated for an additional 10 cycles. Figure 4.81 illustrates the response of an asphalt binder using the creep-recovery test.

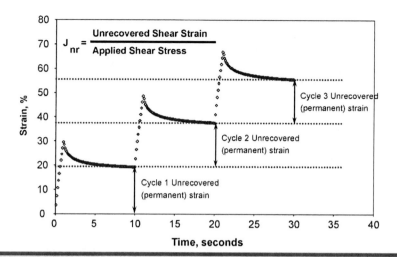

Figure 4.82 Nonrecoverable strain

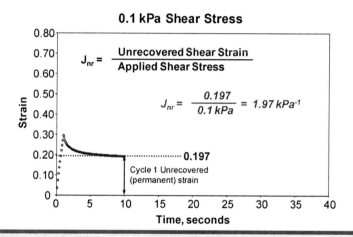

Figure 4.83 Calculating J_{nr}

The material response in the MSCR test is significantly different than the response in the existing PG tests. In the PG system, the high temperature parameter, $G^*/\sin \delta$, is measured by applying an oscillating load to the asphalt binder at relatively low shear strain. In the MSCR test, higher levels of stress and strain are applied to the binder, better representing what occurs in an actual pavement.

In the MSCR test, two separate parameters can be determined—nonrecoverable creep compliance (J_{nr}) and percentage of recovery (MSCR Recovery) during each loading cycle. Values are reported as the average of ten loading cycles at each shear stress level. Nonrecoverable creep compliance, J_{nr}, is determined as shown in Figures 4.82 and 4.83. This parameter has been shown to be better correlated with rutting potential than $G^*/\sin \delta$.

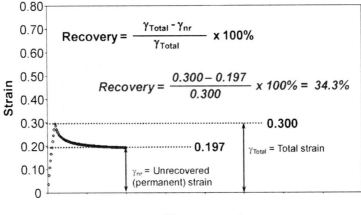

Figure 4.84 Calculating MSCR recovery

Unlike the AASHTO M 320 system, the test temperature used for the MSCR test is selected based on actual high pavement temperatures with no grade bumping. Thus in the previous grade bumping discussion, the MSCR test would be performed on an asphalt binder at a high temperature of 58°C regardless of the traffic speed and loading. In the MSCR specification, AASHTO MP 19, higher traffic loading is accounted for by increasing the stiffness (reducing the compliance) required for the asphalt binder at the grade temperature. For standard traffic loading, J_{nr} (determined at 3.2 kiloPascal shear stress) is required to have a maximum value of 4.0 kiloPascal^{-1}. As traffic increases to heavy and very heavy loading, the J_{nr} of the asphalt binder needs to be lower, requiring a maximum value of 2.0 and 1.0 kiloPascal^{-1}, respectively.

While the main requirement for J_{nr} is determined at 3.2 kiloPascal shear stress, the data determined at 0.1 kiloPascal shear stress is also important. To minimize concerns that some asphalt binders may be overly sensitive to changes in shear stress, AASHTO MP19 maintains a requirement that the difference in J_{nr} values between 0.1 kiloPascal and 3.2 kiloPascal shear stress should not exceed a ratio of 0.75. The calculation is shown below:

$$J_{nr} \text{ Ratio} = \frac{J_{nr,3.2} - J_{nr,0.1}}{J_{nr,0.1}}$$

Although it is not part of the specification, in addition to determining J_{nr} the MSCR test can be used to determine the amount of recovery in an asphalt binder during the creep-recovery testing. Recovery is calculated as shown in Figure 4.84. The MSCR Recovery may be used in combination with J_{nr} to indicate whether an asphalt binder has a sufficient elastic component. This is discussed further in Chapter 4 "Asphalt Modification."

CHAPTER 5

Asphalt emulsions and cutbacks

To use an asphalt cement for construction purposes, it must be liquefied before it can be pumped through pipes, mixed with aggregate, or sprayed through nozzles. Usually the liquification of asphalt cements occurs when the material is heated to elevated temperatures (e.g., 275°F or 135°C). When asphalt cement cools, it becomes a semi-solid cementing material. In some cases it is impractical to use asphalt at such high temperatures. In these instances the asphalt is liquefied by other means to make it workable at ambient or near-ambient temperatures.

To liquefy asphalt for construction operations without heating to high temperatures, the asphalt binder can be diluted (cut) with selected petroleum solvents or emulsified with an emulsifying agent and water. Asphalt binders produced by these methods are known as cutback and emulsified asphalts, respectively.

Once the cutback or emulsified asphalt is used in liquid form during the construction process, the solvent (in cutback asphalt) or water (in emulsified asphalt) will evaporate, leaving the asphalt cement residue to perform its function. Note that the evaporation of petroleum solvents into the atmosphere may be restricted or prohibited by environmental regulations. As a result, the use of cutback asphalts may not be as prevalent as emulsified asphalts.

The process for making emulsified and cutback asphalt materials is shown in the petroleum asphalt flowchart (Figure 1.4). In each case, asphalt cement is the base material that is liquefied by either cutting the asphalt with a petroleum solvent or producing an emulsion of asphalt cement, water, and emulsifying agents.

- **Emulsified asphalts: history and uses**
    Composition
    Types and classifications
    Emulsion components and production
    Breaking and Curing
    Variables affecting emulsion quality

- **Storing asphalt emulsions**
- **Handling asphalt emulsions**
- **Sampling asphalt emulsions**
- **Tests for emulsified asphalts**
    Saybolt Furol viscosity
    Storage stability and settlement
    Classification
    Demulsibility
    Cement mixing
    Coating ability and water resistance
    Particle charge
    CSS classification (mixing)
    Sieve
    Density
    Breaking Index
    Distillation
    Tests on residue
- **Specifications for emulsified asphalts**
- **Cutbacks: Introduction**
    Types and classifications of cutback asphalts
- **Tests for cutback asphalts**
    Kinematic viscosity
    Tag Open Cup Flash Point
    Water
    Distillation
    Tests on residue
- **Specifications for cutback asphalts**

Emulsified asphalts: history and uses

Asphalt emulsions were first developed in the early 1900s and came into general use in pavement applications in the 1920s, primarily in spray applications and as dust palliatives. Growth in the use of emulsified asphalts was relatively slow, limited by the type of emulsions available and a lack of knowledge as to how they should be used. From 1930 to the mid-1950s, there was a steady increase in the use of asphalt emulsions. However, after World War II, roadway designers began to curtail the use of asphalt emulsions on roads in favor of hot mix asphalt (asphalt cement), due to the dramatic increase in traffic loads and volume.

Although the use of asphalt cement has increased greatly since the mid-1950s, the combined use of other asphalt products has remained nearly constant. Nevertheless, there has been a steady rise in the volume of asphalt emulsions used. Continuing development of new types and grades of asphalt emulsions and improved construction equipment and practices now provide the pavement manager with a wide range of choices for use as shown in Table 5.1.

There are many reasons for using asphalt emulsions. Some of them include:
- Asphalt emulsions can be used in preventative maintenance to increase the service life of slightly distressed existing pavements.
- Asphalt emulsion does not require a petroleum solvent to make it liquid and, in most cases, can be used without additional heat. Both of these factors contribute to energy savings.
- Asphalt emulsion may be used when there is concern about reducing atmospheric pollution. There are little to no petroleum distillates used or additional heat needed to make the emulsion liquid, so it produces little to no hydrocarbon emissions.

Table 5.1
Major uses of asphalt emulsions

Surface Treatments	Asphalt Recycling	Other Applications
Scrub Seal	Cold in-place	Cold Mix Paving
Chip Seal	Full Depth	Maintenance Patch
Fog Seal	Hot in-place	Bond (Tack) Coat
Sand Seal	Central Plant Mix	Dust Palliative
Slurry Seal		Prime Coat
Micro Surfacing		Crack Fill
Cape Seal		Protective Coating Stabilization (soil and base)

- Certain types of asphalt emulsions can be used to coat damp aggregate surfaces. This reduces fuel requirements for heating and drying aggregates.
- A variety of emulsion types are available. New formulations and laboratory procedures have been developed to satisfy design and construction requirements.
- Cold emulsions can be used at remote sites.
- Transportation budgets can be stretched by using surface treatments on low and medium volume roads.

Composition

Emulsification is a process through the use of certain mechanical and chemical processes, that allows two or more materials to combine that will not mix under normal conditions. Examples of emulsions include such products as hair dyes, paints, and ice cream. An asphalt emulsion consists of three basic components: asphalt, water, and an emulsifying agent. Other additives—such as stabilizers, coating enhancers, solvents, antistripping additives, and break-control agents—may also be used depending on the formulation and use of the asphalt emulsion. Use of these three components in the proper proportion, and with the proper production, will result in an emulsified asphalt that can be used in the applications described in Table 5.1.

The objective of the production of an asphalt emulsion is to make a stable dispersion of the asphalt cement in water, so that it can be pumped, held in prolonged storage, and mixed. A successfully produced emulsion should "break" quickly after contact with aggregate in a mixer or after spraying on the roadbed ("breaking" is the separation of the water from the asphalt). Upon curing, the residual asphalt retains all of the adhesion, durability, and water-resistance of the asphalt cement from which it was produced. Oftentimes, these properties can be enhanced through proper emulsion formulation.

Types and classifications

By proper selection of emulsifying agents and other manufacturing controls, emulsified asphalts are produced in several types and grades. By choice of emulsifying agent, the emulsified asphalt may be one of the following types:
- Anionic—Asphalt globules are electronegatively charged.
- Cationic—Asphalt globules are electropositively charged.
- Nonionic—Asphalt globules are neutral (have no charge).

Types are determined based on the basic laws of electricity—like charges repel and unlike charges attract. When two poles—an anode and a cathode—are immersed in a liquid and a current is passed through, the anode becomes positively charged and the cathode negatively charged. When this current is passed through an asphalt emulsion, the negatively charged particles will migrate to the anode, hence referred to as an anionic emulsion. Conversely, positively charged asphalt particles will migrate to the cathode and the emulsion is known as cationic. Non-ionic particles are neutral and will not migrate to either pole.

In practice, the first two types have ordinarily been used in roadway construction and maintenance. Non-ionic emulsions, however, have become more widely used as emulsion technology has advanced.

Emulsions are further classified on the basis of how quickly the asphalt droplets will revert to their original state as an asphalt cement. This process is called "coalescence." It is closely related to the speed with which an emulsion will become unstable and break after contact with the surface of an aggregate. Asphalt emulsions are classified as RS, MS, SS, and QS depending on the relative rate of coalescence. RS grades are "rapid setting" and have little or no ability to mix with an aggregate. MS grades are "medium setting" and have the ability to mix with coarse aggregates and clean, fine aggregates. SS and QS grades are "slow setting" and "quick setting," respectively, and have the ability to mix with fine aggregates. QS grades are expected to break more quickly than SS grades (but still more slowly than RS and MS grades).

In addition to the coalescence classification, asphalt emulsions are identified by their particle charge. A "C" in front of the emulsion grade indicates that it is a cationic emulsion. Thus a CSS grade is a slow-setting cationic emulsion. If there is no letter designation before the grade, then it is simply an anionic emulsion.

For some anionic asphalt emulsions, there may be an "HF" designation before the setting classification. These are "High Float" emulsions that have a gel quality imparted by chemical modification that permit a thicker asphalt coating on aggregate particles and prevent draindown of asphalt from the coated aggregate. The high float process also reduces the temperature susceptibility of the residual asphalt binder.

Asphalt emulsions are also identified by a series of numbers and/or letters that follow the coalescence classification. The numbers simply indicate the relative viscosity of the asphalt emulsion, usually designated as a "1" or a "2." Emulsions with a "2" following their classification will have a higher viscosity than emulsions with a "1." Note that these relative indications of viscosity apply to the asphalt emulsion, not the base asphalt used to produce the emulsion. If a harder base asphalt cement is used in the emulsion, it is indicated by an "h" following the number.

Other letter designations may follow the numerical designation of the emulsion viscosity. Typically, these additional letter designations—usually P, S, or L—are indicators that the emulsion has been polymer-modified.

By variations in materials and manufacture, emulsified asphalts are made in several grades as shown in Table 5.2.

Table 5.2
Standard emulsified asphalt grades

	Anionic	Cationic	Cationic, Polymer-Mod.
	ASTM D 977 AASHTO M 140	ASTM D 2397 AASHTO M 208	AASHTO M 316
Rapid Setting (RS)	RS-1	CRS-1	
	RS-2	CRS-2	CRS-2P
	HFRS-2		CRS-2L
Medium Setting (MS)	MS-1		
	MS-2	CMS-2	CMS-2P
	MS-2h	CMS-2h	
	HFMS-1		
	HFMS-2		
	HFMS-2h		
	HFMS-2s		
Slow Setting (SS)	SS-1	CSS-1	
	SS-1h	CSS-1h	
Quick Setting (QS)	QS-1h	CQS-1h	

Note that the list in Table 5.2 is not all-inclusive. For example, micro surfacing applications often use an asphalt emulsion identified as CSS-1hP. Users may also have their own specifications for a particular asphalt emulsion product that do not fit in as an ASTM or AASHTO standard grade.

Besides the fact that Table 5.2 is not all-inclusive, it is highly unlikely that a producer would carry an inventory of all the grades listed in the table. Users should check local supplies before specifying a specific grade of emulsified asphalt.

Emulsion components and production

There are three main components in an asphalt emulsion: asphalt, water, and emulsifying agent. Additives may also be part of the emulsion.

Asphalt

Asphalt cement is the basic ingredient in an asphalt emulsion, making up 50–75 percent of the total emulsion. Asphalt chemistry is a complex subject, and there is no need to examine all the properties of asphalt cement. Some properties of an asphalt cement do significantly affect the finished emulsion, but there is not an exact correlation between these properties and the ease with which the asphalt can be emulsified. Although hardness of base asphalt cements used in producing an asphalt emulsion vary, most often the base asphalt will have penetration between 40 and 250 decimillimeter. Occasionally climatic conditions will require a harder or softer base asphalt cement. Regardless, chemical compatibility between the emulsifying agent and the asphalt cement is essential for production of a stable emulsion.

The complex interaction of the different molecules in an asphalt emulsion makes it nearly impossible to accurately predict the behavior of an asphalt to be emulsified. As such, each emulsion manufacturer has its own formulations and production techniques, with particular attention paid to quality control during production. These formulations and techniques have been developed to achieve optimum results with the materials that are used.

Water

Water is another basic ingredient in an asphalt emulsion, and its contribution to the performance of the finished emulsion product cannot be minimized. Water contains minerals that may affect the production of a stable asphalt emulsion. Thus, water that might be suitable for drinking might not be suitable to produce a stable asphalt emulsion.

Natural, unprocessed water might also be unsuitable for emulsion production because of impurities, either in solution or colloidal suspension. In particular, asphalt emulsion technologists are concerned about the presence of calcium or magnesium ions when producing anionic emulsions. The types of emulsifying agents used to make anionic emulsions react with calcium and magnesium ions to produce insoluble salts that destabilize the emulsion. On the other hand, these same calcium and magnesium ions can actually benefit the production of a cationic emulsion.

In addition to calcium and magnesium ions, particulate matter in the water can lead to destabilization of a cationic emulsion—principally because the (usually) negatively charged particulates react with the positively charged emulsifying agents. This can lead to an imbalance in the emulsion components that can adversely affect performance or cause premature breaking.

Emulsifying agents

Although asphalt cement and water are both key elements in the production of an asphalt emulsion, it is the chemical used as the emulsifying agent that most affects the properties of the final asphalt emulsion. The emulsifier is a surface-active agent, or surfactant, that keeps the asphalt droplets in stable suspension and controls the breaking time. It is the determining factor in the classification of the emulsion as being anionic, cationic, or nonionic.

The most common anionic emulsifiers are fatty acids—wood product derivatives such as tall oils, rosins, and lignins. Anionic emulsifiers are saponified (turned into soap) by reacting with sodium hydroxide or potassium hydroxide.

The most common cationic emulsifiers are fatty amines—diamines, imidazolines, and amidoamines. Cationic emulsifiers are sometimes saponified by reacting with acid, usually hydrochloric. Fatty quarternary ammonium salts are also used to produce cationic emulsions. These are water-soluble salts that may not require the addition of acid to produce a stable cationic asphalt emulsion.

By broad definition, surfactants are water-soluble substances whose presence in solution markedly changes the properties of the water and the surfaces they contact. Surfactants are categorized by the way that they dissociate or ionize in water. Structurally, they possess a molecular balance of a long lipophilic (oil-loving) hydrocarbon tail and a polar, hydrophilic (water-loving) head. Surfactants are adsorbed at the interface between liquids and gases or liquid and solid phases. They tend to concentrate at the interface such that the hydrophilic groups orient themselves towards the more polar phase and the lipophilic groups towards the less polar phase. The surfactant molecule or ion acts as a bridge between two phases. As in the classification of asphalt emulsions, surfactants may be anionic, cationic, or nonionic.

The emulsifying agent is the most important component in an asphalt emulsion formulation. To be an effective emulsifier, a surfactant must be water-soluble and have the right balance between hydrophilic and lipophilic properties. The emulsifier—used in combination with an acceptable asphalt, good quality water, and proper mechanical input—is the major factor in emulsification, emulsion stability, and field performance.

Production

The basic piece of equipment needed to produce an asphalt emulsion is a high-speed, high-shear mechanical device—typically a colloid mill—to divide the asphalt into tiny droplets. Also needed are an emulsifier solution tank, heated asphalt tank, pumps, and flow-metering gauges. A typical manufacturing layout is shown in Figure 5.1.

The colloid mill has a high speed rotor that revolves between 1000 and 6000 RPM with mill clearance settings typically between 0.01 to 0.02 inches (0.25–0.50 millimeter). Usually asphalt emulsions have droplet sizes that are smaller than the diameter of a human hair, approximately 0.001 to 0.010 millimeter. Asphalt droplet sizes depend upon the mechanical energy density imparted by the mill. Particle size analyzers may be used by manufacturers to characterize emulsion quality.

Separate pumps are used to deliver asphalt and emulsifier solution to the colloid mill. Because the emulsifier solution can be corrosive, it may be necessary to use equipment made with corrosion-resistant materials.

Prior to the emulsification process, asphalt and water are heated separately to the optimum temperature. This temperature may vary depending on the properties of the asphalt and the compatibility between the asphalt and the emulsifying agent. The asphalt is fed into the colloid mill where it is divided into tiny droplets. At the same time, water containing the emulsifying agent is also fed into the colloid mill. The temperature of the emulsion leaving the mill must be below the boiling point of water. After production, the emulsion is pumped into bulk storage tanks, which may be equipped with mechanical agitation to keep the emulsion uniformly blended. Nitrogen blankets may also be used in storage to prevent skinning or breaking of the emulsion.

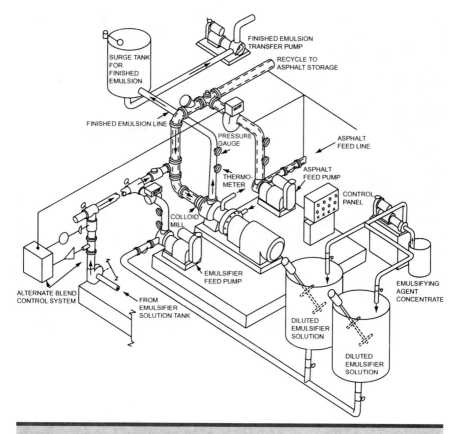

Figure 5.1 Schematic of emulsifying plant

The method of adding the emulsifier to the water varies according to the manufacturer's procedure. To be water-soluble, amines may be mixed and reacted with an acid and fatty acids may be mixed and reacted with an alkali. Emulsifier mixing is typically conducted in a batch mixing tank where the emulsifier is introduced into warm water which may contain either acid or alkali. It is agitated until the emulsifier is completely dissolved.

To produce an acceptable emulsion, the asphalt and emulsifier solution must be proportioned accurately—a process usually done with flow meters. Proportioning can also be accomplished by monitoring the temperatures of the asphalt and emulsifier solution entering the mill as well as monitoring the discharge temperature.

Asphalt particle size is a vital factor in producing a stable asphalt emulsion. On average, 20 percent of the asphalt particles in an emulsion are smaller than 0.001 millimeter, 57 percent are between 0.001 and 0.005 millimeter, and 23 percent are between 0.005 and 0.010 millimeter. These microscopic asphalt droplets (Figure 5.2) are dispersed in a solution of water and surfactant. The surfactant causes a

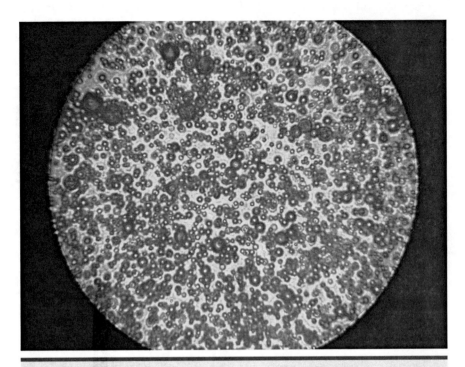

Figure 5.2 Asphalt particles in emulsion

change in surface tension at the contact area between the asphalt droplets and the surrounding water, which allows the asphalt to remain in a suspended state. Since the asphalt particles all received a similar electrical charge from the emulsifying agent, they repel each other, which aids in keeping them suspended in the emulsion.

Breaking and curing

If an asphalt emulsion is to perform its ultimate function as a binder, then the water phase must separate from the asphalt phase and evaporate. This separation is referred to as "breaking." Emulsions are formulated to break according to their intended use. Emulsions may break chemically or by evaporation, depending on the coalescence time that is desired. SS emulsion grades typically break by evaporation. For MS and RS emulsion grades the breaking mechanism is mostly chemical.

The specific type and concentration of the emulsifying agent primarily controls the rate of breaking. However, other factors may also play an important role in the breaking of an emulsion. In order to achieve optimum results, it is necessary to control these other factors to meet the specific requirements of the field use of the asphalt emulsion.

Curing is the development of the mechanical properties of the asphalt cement after it is in-place. For this to happen, the water must completely evaporate from the emulsion and the asphalt particles need to coalesce and bond to the intended surface. Water is removed by both evaporation and absorption. With MS and SS emulsion grades for mixes, the use of slightly damp aggregates facilitates the mixing and coating process. The development in strength in the SS emulsion grades depends mainly on the evaporation and absorption of the water by the aggregates.

Some asphalt emulsions may contain petroleum solvents to aid in the mixing and coating process. The type and quantity of solvent affect the speed of the curing process.

Factors affecting breaking and curing

Some of the factors affecting breaking and curing include:
- Water absorption: A rough-textured, porous aggregate surface speeds the setting time by absorbing water from the emulsion.
- Aggregate moisture content: While damp aggregate may facilitate coating, it slows the curing process by increasing the amount of time needed for evaporation.
- Weather conditions: Temperature, humidity, and wind velocity affect water evaporation rate, emulsifier migration, and water release characteristics. Usually breaking occurs more quickly at warmer temperatures and lower levels of relative humidity. In hot weather, skin formation might form on chip seals, trapping water and delaying curing.
- Mechanical forces: Roller pressure (slow moving traffic) forces the water from the mix and helps attain mix cohesion, cure, and stability.
- Surface area: Greater aggregate surface area, particularly excessive fines or dirty aggregate, accelerates the breaking of the emulsion.
- Surface chemistry: The intensity of the aggregate surface charge and emulsifier charge can impact setting rate, particularly for cationic emulsions. Calcium and magnesium ions on the aggregate surface can react with and destabilize certain anionic emulsifiers, accelerating setting.
- Emulsion and aggregate temperature: Breaking is accelerated when emulsion temperatures are high. This is particularly evident in microsurfacing.
- Type and amount of emulsifier: The surfactant determines the breaking characteristics of the emulsion.

Variables affecting emulsion quality

Many factors can affect the production, storage, and performance of an asphalt emulsion, including:
- Chemical properties of the base asphalt cement
- Hardness and quantity of the base asphalt cement
- Asphalt particle size in the emulsion
- Type and concentration of the emulsifying agent
- Manufacturing conditions such as temperature, pressure, and shear (time and rate)

- Ionic charge on the emulsion particles
- Order of addition of the ingredients
- Type of equipment used in manufacturing, storage, and application of the emulsion
- Properties of the emulsifying agent
- Addition of chemical modifiers or polymers
- Water quality (hardness)

Storing asphalt emulsions

As an asphalt emulsion is a dispersion of fine asphalt droplets in water, it has special storage requirements. Storage tanks should be insulated to protect them from freezing as well as to promote an efficient use of heat. Since a skin of asphalt can form on the surface of emulsions when they are exposed to air, it is best to use tall, vertical storage tanks that minimize the amount of surface area that is exposed to air. If horizontal tanks are used (often for short-term field storage), skin formation can be minimized by keeping the tanks as full as practical to minimize the surface area exposed to air. Nitrogen blankets may also be used to minimize the possibility of skin formation.

Side-entering propellers located approximately 3 feet (1 meter) from the tank bottom may be used to prevent surface skin formation. These propellers should have a large diameter and should be slowly turning (approximately 60 RPM) to gently circulate the material. Propellers should only be operated when there is sufficient emulsion for proper mixing.

Table 5.3 provides some storage guidelines for asphalt emulsions.

Table 5.3
Asphalt storage guidelines

DOs	DON'Ts
Store between 50°F and 185°F (10°C and 85°C) depending on the use and product (see Table 5.4)	Heat above 185°F (85°C). This could result in rapid evaporation of the water, changing the characteristics of the emulsion.
	Allow the emulsion to freeze. This breaks the emulsion.
	Allow the temperature of the heating surface to exceed 212°F (100°C). This could cause premature breakdown of the emulsion on the heating surface.
	Use forced air to agitate the emulsion. It could cause the emulsion to break.
	Use excessive agitation by mixing or pumping.

Table 5.4
Storage temperatures

Grade	Storage Temperature Minimum	Storage Temperature Maximum
QS-1h, CQS-1h, Micro Surfacing Emulsion	50°F (10°C)	125°F (50°C)
RS-2, CRS-1, CRS-2, HFRS-2, MS-2, MS-2h, CMS-2, CMS-2h, HFMS-2, HFMS-2h	125°F (50°C)	185°F (85°C)
RS-1, MS-1, SS-1, SS-1h, CSS-1, CSS-1h	50°F (10°C)	140°F (60°C)

Handling asphalt emulsions

As with storage, proper handling of asphalt emulsions is also very important. Table 5.5 provides a list of "Dos" and "Don'ts."

Table 5.5
Handling asphalt emulsions

DOs	DON'Ts
Provide adequate ventilation. Avoid exposure to fumes, vapors, and mist.	Mix different classes, types, and grades in tanks, transports, or distributors.
Agitate gently when heating to eliminate or reduce skin formation.	Load into tanks, transports, or distributors containing remains of incompatible materials.
Protect pumps, valves, and lines from freezing.	Apply severe heat to pump packing glands or pump casings.
Clear out lines and leave drain plugs open when not in use.	Dilute RS grades with water. MS and SS grades may be diluted, but always add water slowly to the emulsion. Never add the asphalt emulsion to water when diluting.
Use pumps with proper clearances.	
Warm the pump to ~150°F (65°C) to facilitate startup.	Subject the asphalt emulsion, or air above it, to open flame, heat, or strong oxidants.
When diluting, check the compatibility of the water with the emulsion.	
If possible, use warm water for diluting, and always add the water slowly to the emulsion. Never add the asphalt emulsion to water when diluting.	
Avoid repeated pumping and recirculating as viscosity may drop, and air may be entrained, causing the emulsion to be unstable.	
Position inlet pipes and return lines at the bottom of tanks to prevent foaming.	
Pump from the bottom of the tank to minimize contamination from skin formation (if present).	
Haul emulsion in transports with baffle plates to prevent sloshing.	
Agitate emulsions slowly that have been in prolonged storage.	
Consult the supplier(s) before mixing different emulsions of the same grade to verify that they are compatible.	

Sampling asphalt emulsions

The purpose of any sampling method is to obtain samples that will show the true nature and condition of the material. The procedure is described in "Standard Methods of Sampling Bituminous Materials" (ASTM D 140, AASHTO T 40).

Appropriate sampling containers include: (1) plastic, wide-mouth jars or bottles; (2) wide-mouth, plastic-lined cans with lined screw caps; or (3) plastic-lined, triple-seal, friction-top cans. Generally 1-gallon (4-liter) sample containers are preferred.

Whenever practical, the asphalt emulsion shall be sampled at the point of manufacture or storage. Samples may also be taken from the shipment immediately upon delivery. Three samples shall be taken and sent as soon as possible to the laboratory for testing. Samples should not be exposed to freezing temperatures or extreme heat.

Since test results are highly dependent on proper sampling procedure, extra care is required to ensure that the samples obtained are truly representative of the material being sampled. Sampling should only be conducted by trained, authorized personnel. Some sampling, protection, and preservation precautions are as follows:

- Sample containers shall be new. They shall not be washed or rinsed. If they contain evidence of solder flux, or if they are not clean and dry, they should not be used. Top and container shall fit together tightly.
- Care shall be taken to prevent the samples from becoming contaminated. The sample container shall not be submerged in solvent, nor shall it be wiped with a solvent-saturated cloth. Any residual material on the outside of the sample container shall be wiped with a clean, dry cloth immediately after the container is sealed and removed from the sampling device.
- The sample shall not be transferred to another container.
- The filled sample container shall be tightly and positively sealed immediately after the sample is taken.
- Immediately after filling, sealing, and cleaning, the sample container shall be properly marked for identification with a permanent marker, label, or sample tag on the side of the container, not the lid.
- Emulsion samples shall be packaged, labeled, and protected from freezing during shipment.
- Emulsion samples should be packaged and shipped to the testing laboratory the same day they are taken. The containers should be tightly sealed and packed in protective material to reduce the probability of damage during shipment.
- Samples should be tested within two weeks of the shipping date.

- Samples that will be submitted to an agency or outside testing lab should be identified with the following information:
    - Shipper's name and bill of lading (or loading slip number)
    - Sampling date
    - Sampler's name
    - Product grade
    - Project identification
    - Other important information as needed

Tests for emulsified asphalts

Proper interpretation of laboratory test results can greatly aid in determining the traits of an asphalt emulsion. Some of these tests are designed to measure performance qualities, while others are designed to measure composition, consistency, and stability. These tests are normally conducted to: (1) provide data for specification requirements; (2) assess the quality and uniformity of the product during manufacturing and use; and (3) predict the handling, storage, and field performance properties of the material.

A review of asphalt emulsion specifications across the United States reveals that there are a wide variety of requirements, many directly related to the emulsions produced by specific manufacturers. Rather than discuss the myriad of tests and specifications currently in use, the following will focus on the ASTM and AASHTO standards which represent the majority of test methods used to characterize asphalt emulsions.

Saybolt Furol viscosity

The Saybolt Furol viscosity test is used to measure the consistency (rate of flow) of emulsified asphalts. As a matter of testing convenience, and also to achieve suitable accuracy, the test is performed at one of two temperatures, 77°F or 122°F (25°C or 50°C), depending on the viscosity characteristics of the specific type and grade of asphalt emulsion.

For testing at 77°F (25°C), a sample is carefully stirred and conditioned for 30 minutes at the test temperature. The sample is then poured through a 0.85-millimeter sieve or strainer into a standard Saybolt Furol viscometer with a stopper placed in the orifice. The test is started by first withdrawing the stopper and determining the time required for 60 milliliters of emulsion to flow through the orifice (Figure 5.3). The time interval is called Saybolt Furol viscosity and is measured in seconds. The more viscous the material, the greater the length of time required for a given volume to flow through the orifice. Thus, a higher Saybolt Furol time is indicative of a higher sample viscosity.

For testing at 122°F (50°C), the sample is first heated to 122±5°F (50±3°C) and then poured through the strainer into the viscometer and brought to test temperature before the stopper is removed and the flow is timed, as already described.

The Saybolt Furol viscosity test is suitable for testing emulsified asphalts that have a minimum flow time of 20 seconds.

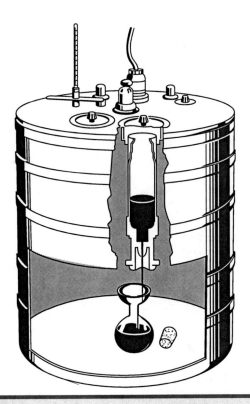

Figure 5.3 Saybolt Furol viscosity

Storage stability and settlement

The purpose of the storage stability and settlement tests is to detect the tendency of asphalt globules to "settle out" during storage of emulsified asphalts. The user is provided with an element of protection against separation of asphalt and water in unstable asphalt emulsions that may be stored for a period of time.

Either test is conducted by placing a 500-milliliter sample in a graduated cylinder, which is then sealed with a stopper and allowed to stand undisturbed. For the storage stability test, the sample must remain undisturbed at 77±5°F (25±3°C) for 24 hours. For the settlement test, the sample must remain undisturbed at 77±5°F (25±3°C) for 5 days. At the end of the storage period, small samples are then taken from the top and the bottom parts of the cylinder. Each sample is placed in an individual beaker and weighed. The samples are then heated to 325°F (163°C) for a total of 3 hours, allowed to cool to room temperature, and weighed. The weights obtained provide the basis for determining the difference, if any, between asphalt cement content in the upper and lower portions of the graduated cylinder, thus providing a measure of the settling of the asphalt globules from the emulsion.

Classification

The classification test is a procedure for distinguishing rapid-setting cationic emulsified asphalt from other types by their failure to coat a sand-cement mixture.

The procedure requires that a washed and dried sample of Ottawa Sand (465 grams) be mixed in a beaker with a 35-gram sample of emulsified asphalt and thoroughly mixed for 2 minutes. At the end of the mixing period, any excess emulsion is drained off and the mix is placed on absorbent paper for visual examination. An excess of uncoated area (compared with the coated area of sand particles) is considered a passing rating for rapid-setting cationic emulsions.

Demulsibility

The demulsibility test is used to identify rapid-setting emulsified asphalts by indicating the relative rate at which the colloidal asphalt globules will break when spread in thin films on soil or aggregate. Calcium chloride causes the minute asphalt globules present in these emulsified asphalts to coalesce. In the test, a solution of calcium chloride and water is thoroughly mixed with emulsified asphalt and then decanted over a sieve to determine how much the asphalt globules coalesce.

In rapid-setting (RS) emulsions tests, a very weak solution of calcium chloride and water is used. Specifications prescribe the concentration of the solution and the minimum amount of asphalt to be retained on the sieve. A high degree of "demulsibility" indicates an RS emulsion, which is expected to break almost immediately upon contact with the aggregate on which it is applied.

When cationic rapid-setting (CRS) emulsions are tested, a solution of sodium dioctyl sulfosuccinate in water is used in place of a calcium chloride solution.

Cement mixing

The cement-mixing test is used instead of the demulsibility test for slow-setting (SS) grades of emulsified asphalt. This test is specified for both the anionic and cationic types to ensure that the products are substantially immune from a rapid coalescence of asphalt particles in contact with fine-graded soils or dusty aggregates.

The cement-mixing test is performed by stirring together 100 milliliters of emulsion—which has been diluted with water to 55 percent residue—with 50 grams of high-early-strength Portland cement (Type III). After these ingredients are stirred for 1 minute, 150 milliliters of water is added and the mixture is stirred for an additional 3 minutes. The mixture is then washed over a 1.40-millimeter (No. 14) sieve and the percent of material retained on the sieve is determined. A low amount of retained weight (i.e., particles that have "broken") is desirable for SS grades.

Coating ability and water resistance

This test has a threefold purpose. It determines the ability of an asphalt emulsion to coat aggregate thoroughly, to withstand mixing action while remaining as a film on the aggregates, and to resist the washing action of water after completion of mixing. The test is primarily intended to identify medium-setting asphalt emulsions suitable for mixing with coarse-graded calcareous aggregates. Other aggregates may be used in the test if calcium carbonate is omitted throughout the method. This test is not adaptable to rapid-setting or slow-setting asphalt emulsions.

The test is conducted by coating 461 grams of a standard reference aggregate with 4 grams of calcium carbonate dust. The combined aggregate is then mixed with 35 grams of emulsified asphalt for 5 minutes. About half of the mixture is then placed on absorbent paper for a visual inspection of the surface area of aggregate coated by the emulsified asphalt. The remainder of the mixture is sprayed with water and rinsed until the rinse water runs clear. This material is then placed on absorbent paper and inspected for coating retention.

The procedure is then repeated using wet aggregate. Visual inspections are made as previously described for the dry aggregate tests.

Particle charge

The particle charge test is used to identify cationic asphalt emulsions. The test is conducted by heating a sample of emulsion to 50°C and pouring it into a 250-milliliter beaker. A positive electrode (anode) and a negative electrode (cathode) are then immersed in the sample and connected to a controlled direct-current electrical source providing a current of at least 8 milliamperes. After 30 minutes, or after the current has dropped to 2 milliamperes, the 2 electrodes are examined. An asphalt deposit on the cathode that is clearly discernible (compared with the anode) indicates that the sample is a cationic-emulsified asphalt.

Note that this test may not produce conclusive results for CSS grades. If the test is inconclusive, then the pH of the sample should be determined.

CSS classification (mixing)

If an inconclusive result is obtained in the particle charge test for a CSS emulsion, then a classification (mixing) test may be conducted. In this test, a weighed amount of washed and dried silica sand is mixed with a weighed amount of cationic slow-setting (CSS) asphalt emulsion and mixed until the aggregate is completely coated. The amount of emulsion in the mix should be 5 percent by total weight of sand. The mix is cured for 24 hours and then placed in a beaker of boiling, distilled water. After 10 minutes, the sample is placed on a level surface and the coating is observed. If the coating is in excess of 50 percent of the total mix, it is considered a positive identification of a CSS asphalt emulsion.

Sieve

The sieve test complements the settlement test and has a somewhat similar purpose. It is used to determine quantitatively the percent of asphalt cement present in the form of pieces, strings, or relatively large globules. Such particles of asphalt might clog equipment and tend to provide non-uniform coatings of asphalt on aggregate particles. This non-uniformity might not be detected by the settlement test, which is of value in this regard only when there is a sufficient difference in the specific gravity of asphalt and water to allow settlement.

The test is conducted by bringing 1000 grams of asphalt emulsion to the proper test temperature depending on its Saybolt Furol viscosity. Emulsified asphalts having a high viscosity—greater than 100 seconds at 77°F (25°C)—or any emulsions required to be tested at 122°F (50°C) are tested at 122°F (50°C). All other samples are tested at ambient temperature. The sample is then poured through an 850-μm (No. 20) sieve. For anionic emulsified asphalts, the sieve and retained asphalt are then rinsed with a mild sodium oleate solution. For cationic emulsified asphalts, distilled water is the rinse agent. After rinsing, the sieve and asphalt are dried in an oven and the relative amount of asphalt retained on the sieve is determined. A low percentage of retained material is desirable in this test.

Density

Density is calculated by determining the mass of an asphalt emulsion in a standard measure of known volume. Results are reported to the nearest 0.01 pound/gallon at 77°F (0.01 kilogram/liter at 25°C). The results may be used to calibrate the distributor to ensure that the correct shot rate is used.

Breaking Index

The Breaking Index test is designed to measure the speed of break of rapid-setting asphalt emulsions. Silica sand is added to 100 grams of the asphalt emulsion while stirring under controlled conditions of rate and temperature. The weight of the sand, in grams, needed to cause the emulsion to break (become black and sticky) is referred to as the Breaking Index.

Distillation

The distillation test is used to determine the relative proportion of asphalt cement and water in the asphalt emulsion. Some grades of emulsified asphalt also contain an oil distillate; the distillation test provides information on the amount of this material in the emulsion. Also, the distillation test provides an asphalt cement residue, on which additional tests (penetration, solubility, and ductility) may be made, as previously described for asphalt cement.

As with the sieve test, emulsified asphalts with a high viscosity are heated to 122°F (50°C) before beginning the procedure. In the distillation procedure, a 200-gram (for unmodified, not modified, emulsions) sample of emulsion is placed in an aluminum-alloy still and heated (Figure 5.4). Once the temperature reaches 500°F (260°C), distillation is continued for 15 minutes to produce a smooth, homogeneous residue, which may then be used for further tests.

Rapid- and medium-setting grades of cationic asphalt emulsion may include an oily distillate, the maximum amount of which usually is limited by specifications. The distillate collected in the graduated cylinder includes both oil and water from the asphalt emulsion. Because these two materials separate in the graduated cylinder, the amounts of each can be determined.

Since the asphalt emulsion will likely never experience such an elevated temperature in the field, it should be noted that some residue properties could be altered, such as elastic properties provided by polymer modification. Some agencies have changed the time and temperature at which the distillation test is conducted for these products.

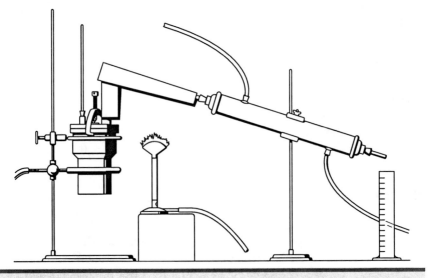

Figure 5.4 Emulsion distillation

One technique that may be considered for polymer-modified emulsions is the evaporation procedure described in ASTM D 6934, "Standard Test Method for Residue by Evaporation of Emulsified Asphalt." In this procedure, 3 beakers, each containing 50 grams of emulsified asphalt, are heated for 2 hours in a 325°F (163°C) oven, then removed, stirred, and replaced in the oven for an additional hour. At the end of the additional hour, the beakers are removed and allowed to cool to room temperature. The beakers are weighed to determine the residue percentage. If the residue is needed for testing, the beakers are placed in the oven until the residue is fluid enough to pass through a 300-millimeter (#50) sieve. The ASTM standard notes that this procedure ". . . tends to give an asphaltic residue lower in penetration and ductility than the distillation test method (D 6997)."

Another technique for obtaining residue from an emulsified asphalt is described in ASTM D 244 as the low temperature vacuum distillation procedure. This procedure is particularly suitable for asphalt emulsion residue properties that may be altered by standard, high-temperature distillation (i.e., polymer-modified emulsified asphalt). In this procedure, a 300-gram sample of emulsion is heated while using an applied vacuum of 88 kiloPascal below atmospheric pressure. The emulsion is heated for 45 minutes until it reaches 275°F (135°C) and is then held at this temperature for 15 minutes. The ASTM standard notes that the low temperature vacuum distillation procedure is ". . . not intended to produce residues equivalent to the D 6997 500°F (260°C) distillation procedure."

Finally, ASTM has recently published a new procedure—D 7497, "Standard Practice for Recovering Residue from Emulsified Asphalt Using Low Temperature Evaporative Technique"—that takes two days to complete, but has a maximum temperature of 140°F (60°C).

Tests on residue

After distillation, the asphalt residue may be tested using standard asphalt binder tests. As of this printing, standard specifications for emulsified asphalt residue still use tests from the viscosity and penetration grading systems rather than from the PG asphalt binder system. These tests may include penetration, ductility, and solubility. Other tests used for modified asphalt emulsions—such as elastic recovery, force ductility, and toughness and tenacity—are discussed later in this manual.

Float test

One additional test is commonly conducted on residue produced from high float (HF) emulsions. The float test uses a semispherical float made from aluminum or aluminum-alloy that has an 11-millimeter diameter opening in the bottom of the float. The residue sample is heated and poured into a brass collar designed to be screwed in-place in the float. After the sample cools in the collar, it is placed in the float and allowed

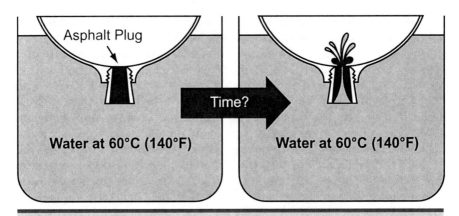

Figure 5.5 Float test

to float in a water bath operating at 140°F (60°C) (Figure 5.5). The time is measured from the instant the float is placed in the bath until water breaks through the asphalt residue in the float, causing it to sink.

Specifications for emulsified asphalts

Emulsified asphalts, which are produced using a variety of asphalt cement bases and setting properties, are manufactured for a number of uses. Specifications for anionic and cationic asphalt emulsions are included in AASHTO and ASTM standards. For anionic emulsified asphalts (ASTM D 977 and AASHTO M 140), there are four standard types available: rapid-setting (RS), medium-setting (MS), slow-setting (SS), and quick-setting (QS). The same setting types also are available for cationic emulsified asphalts (ASTM D 2397 and AASHTO M 208) and are designated CRS, CMS, CSS, and CQS. AASHTO M 316 provides specifications for cationic polymer-modified emulsions.

Five grades of high-float anionic emulsions—designated as HFRS-2, HFMS-1, HFMS-2, HFMS-2h, and HFMS-2s—are also part of the ASTM D 977 specification. There are no corresponding high-float cationic emulsions in standard specifications. High-float emulsions have a specific quality that permits a thicker film coating with a reduced risk of runoff.

Several tests are required to determine specification compliance of emulsified asphalts. All grades of emulsified asphalt have requirements on Saybolt Furol viscosity, storage stability (except for QS grades), sieve, and percentage of residue. The emulsified asphalt residue also has requirements for all grades on penetration, ductility, and solubility.

In addition to the above tests common for all emulsified asphalt grades, all cationic emulsified asphalt grades require a positive result on the particle charge test.

Additional tests, depending on the emulsified asphalt grade, are included in the specifications. For example, the demulsibility test is required only for rapid-setting (RS) grades. The coating ability and water resistance tests are only required for medium-setting (MS) grades. The cement-mixing test is required only for slow-setting (SS) grades.

These tests are listed in Table 5.6. Some of the tests for asphalt cement are also required for testing emulsified asphalt residue.

Table 5.6
Specification tests for emulsified asphalt grades

Test	AASHTO M 140 ASTM D 977						AASHTO M 208 ASTM D 2397			
	RS	HFRS	MS	HFMS	SS	QS	CRS	CMS	CSS	CQS
Viscosity, Saybolt	x	x	x	x	x	x	x	x	x	x
Storage Stability	x	x	x	x	x		x	x	x	
Sieve	x	x	x	x	x	x	x	x	x	x
Distilled Residue %	x	x	x	x	x	x	x	x	x	x
Demulsibility	x	x					x			
Coating Ability and Water Resistance			x	x				x		
Cement Mixing					x				x	
Oil Distillate %				x[1]			x	x		
Particle Charge							x	x	x	x
Residue Tests										
Penetration	x	x	x	x	x	x	x	x	x	x
Ductility	x	x	x	x	x	x	x	x	x	x
Solubility	x	x	x	x	x	x	x	x	x	x
Float Test		x		x						

[1] HFMS-2s only

Cutbacks: Introduction

If an asphalt binder is liquefied by diluting it with selected petroleum solvents, it is known as a cutback asphalt. Once the cutback asphalt is used in liquid form during the construction process, the solvent will evaporate, leaving the asphalt cement residue to perform its function. Note that the evaporation of petroleum solvents into the atmosphere may be restricted or prohibited by environmental regulations. As a result, the use of cutback asphalts may not be as prevalent as emulsified asphalts.

Uses of cutback asphalts

Cutback asphalts can be used with cold aggregates, with a minimum of heat. RC and MC types of cutback asphalt have been widely used in a variety of roadway, airfield, industrial, and specialty uses. Among the more important uses were road-mixing operations, stockpile mixes, and spray applications (such as prime, tack, and seal coats). But because the diluents in cutback asphalts are high-energy products that are lost by evaporation and because of environmental regulations, emulsified asphalts have largely supplanted cutbacks for most of these uses. SC cutback asphalts are used primarily in road-mixing and dust-laying applications. They are also used for stockpile patching mixes, plant-mixing with graded aggregates, and occasionally for priming. Here too, use is declining. It is suggested that local environmental regulations be reviewed before using a cutback asphalt.

Types and classifications of cutback asphalts

Petroleum solvents used for diluting asphalt cement are variously called distillate, diluent, or cutter stock. If the solvent used in making the cutback asphalt is highly volatile, it will quickly escape by evaporation. Solvents of lower volatility evaporate more slowly. Cutback asphalts are divided into the following three types based on the relative speed of evaporation:
- Rapid-curing (RC)—asphalt cement combined with a light diluent of high volatility, generally with a boiling point similar to gasoline or naphtha. Grades include RC-70, RC-250, RC-800, and RC-3000.
- Medium-curing (MC)—asphalt cement combined with a medium diluent of intermediate volatility, generally with a boiling point similar to kerosene. Grades include MC-30, MC-70, MC-250, MC-800, and MC-3000.
- Slow-curing (SC)—asphalt cement combined with oils of low volatility. Grades include SC-70, SC-250, SC-800, and SC-3000.

Slow-curing cutback asphalts may also be referred to as road oils. This term originated in earlier days when asphalt residual oil was used to give unpaved roads a low-cost, all-weather surface.

The degree of liquidity developed in each cutback asphalt depends principally on the proportion of solvent to asphalt cement. To a minor degree, the liquidity of the cutback may be affected by the hardness of the base asphalt from which the cutback is made. The more viscous grades of cutback asphalt are indicated by higher numbers in the grade designation and may require a small amount of heating to make them fluid enough for construction operations.

Tests for cutback asphalts

Kinematic viscosity

The kinematic viscosity test is used as a basis for classifying cutback asphalts into standard grades. The test procedure is similar to the kinematic viscosity test used for asphalt binders, except that the test is conducted in a thermostatically controlled water bath operating at 140°F (60°C).

Tag Open Cup Flash Point

The purpose and significance of the Tag Open Cup Flash Point test is the same as the Cleveland Open Cup Flash Point test described previously. The Tag Open Cup test is used for materials having flash points less than 199°F (93°C). This includes RC and MC grades, but not SC grades, which use the Cleveland Open Cup test.

The Tag Open Cup Flash Point procedure is essentially the same as the Cleveland Open Cup procedure, except that indirect heating is used because of the volatile nature of the diluent used in these products. The Tag Open Cup apparatus is glass, rather than metal, and is heated in a water bath rather than by direct flame.

Cutback asphalts commonly are used at temperatures above their flash point. Some rapid-curing cutbacks may flash as low as 80°F (27°C). The more volatile the solvent the cutback asphalt contains, the more hazardous its use. All these materials present some danger in use and must be handled properly.

Water

Asphalt materials, except for emulsified asphalts, are usually specified to be substantially water free, primarily because water causes foaming when the material is heated, creating a potentially hazardous condition.

To determine the amount of water, if any, in cutback asphalt, a measured volume of asphalt is poured into a glass or metal still and thoroughly mixed with a xylene or high-boiling-range petroleum naphtha. A reflux condenser is attached to the still and discharged into a graduated trap. Heat is applied to the still and any water present is collected in the trap. The percentage of water by volume is then calculated.

Distillation

As previously indicated, cutback asphalts are blends of asphalt cement and suitable diluents. Properties of these materials are of importance in their application and performance.

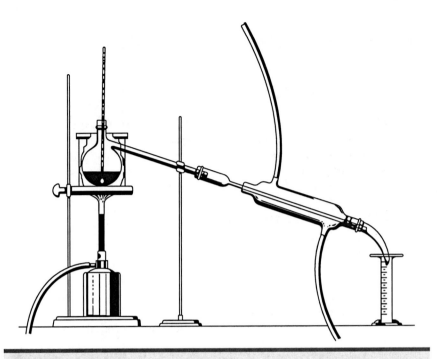

Figure 5.6 Cutback distillation

The asphalt cement and diluent are separated by the distillation test (Figure 5.6) to determine their quantity and for further identification. Approximately 200 milliliters of cutback asphalt is poured into a distillation flask connected to a water-cooled condenser tube. As the flask is slowly heated, the diluent vaporizes in the distillation flask and is again liquefied in the condensing tube. The condensate is drained into a graduated cylinder. The quantity of condensate driven off at several specified temperatures is determined, and this is an indication of the volatility characteristics of the diluent. After reaching 680°F (360°C), the material remaining in the distillation flask is considered asphalt cement residue.

The distillates that evaporate at various temperatures are of little importance for SC cutback asphalts. For these grades, the distillates that evaporate below the specified end point, 680°F (360°C), are largely of an oily nature, meaning that their rate of evaporation in service is quite slow. Therefore, only the total quantity of distillate driven off up to 680°F (360°C) is measured.

Tests on residue

As with emulsified asphalt grades, the asphalt residue remaining after distillation may be tested using standard asphalt binder tests. As of this printing, standard specifications for cutback asphalt residue still use tests from the viscosity and penetration grading systems rather than the PG asphalt binder system. These tests may include viscosity, penetration, ductility, and solubility.

Residue of 100 penetration

This test, described in ASTM D 243, is conducted only on SC cutback grades, and is perhaps the least significant to the consumer of all tests normally used on these materials. Because the in-service rate of curing of an SC cutback asphalt is quite slow, it may or may not reach a penetration of 100 during its service life. The principal value of the test is that it provides a residue upon which standard tests for asphalt binders may be made.

The test is performed by rapidly heating a small sample (100 grams) of SC cutback asphalt from 480°F to 500°F (249°C to 260°C) and maintaining it at this temperature until the operator judges that the material has achieved a penetration of 100 decimillimeters. The weight of asphalt residue remaining is determined and the residue percentage is calculated.

Specifications for cutback asphalts

Cutback asphalts, which are produced using a variety of asphalt cement bases and diluents, are manufactured for a number of uses. AASHTO and ASTM have specifications for three standard types of cutbacks: rapid-curing (RC), medium-curing (MC), and slow-curing (SC). ASTM D 2026, D 2027, and D 2028 provide standard specifications for SC, MC, and RC cutbacks, respectively. In the specifications, the type (RC, MC, or SC) indicates the relative speed of evaporation. The grades (70, 250, 800, and 3000) indicate the minimum allowable kinematic viscosity in centistokes at 140°F (60°C). One additional grade, MC-30, has been used as a special grade for prime-coat applications in some sections of the United States. The maximum allowable viscosity for each grade is twice the minimum allowable value.

The most viscous grades (RC-3000, MC-3000, and SC-3000) are only moderately less viscous than the lowest viscosity-graded asphalt binders. The least viscous grades (RC-70, MC-30, MC-70, and SC-70) may be readily poured at room temperature.

Several tests are required to determine specification compliance of cutback asphalts. These tests are shown in Table 5.7 and are substantially the same for all cutback types and grades. The only exceptions are SC cutback asphalts that use the Cleveland Open Cup (COC) Flash Point test instead of the Tag Open Cup (TOC) Flash Point test, which is used by RC and MC cutback asphalts. SC cutbacks also have an additional test (the 100 Pen Residue test) that is not common to the other grades.

Table 5.7
Specification tests for cutback asphalt grades

Test	Method AASHTO	Method ASTM	Spec[1] RC	Spec[1] MC	Spec[1] SC
Viscosity @ 60°C	T 201	D 2170	x	x	x
Flash Point, TOC	T 79	D 3143	x	x	
Flash Point, COC	T 48	D 92			x
Distillation	T 78	D 402	x	x	x
Residue Tests					
Viscosity @ 60°C	T 202[2]	D 2171[2]	x	x	x
Ductility	T 5	D 113	x	x	x
Solubility	T 44	D 2042	x	x	x
100 Pen Residue		D 243			x
Water %	T 55	D 95	x	x	x

[1] Cutback asphalt specifications are ASTM D 2026, D 2027, and D 2028.
[2] For SC cutback asphalts, substitute T 201 or ASTM D 2170.

CHAPTER 6

Air-blown asphalt

Special properties may be imparted to asphalt binders by blowing air through the topped crude fraction during the latter part of the refining process. The regular distillation process is discontinued at some point while the topped crude is still liquid. The heavy fraction is then put into a converter and air is blown through it while it is maintained at a high temperature. This process is continued until the asphalt has attained desired properties. Often, such asphalts are called oxidized asphalts. However, this is not strictly a true term because not only oxidation but also vaporization, dehydrogenation, condensation, polymerization, and other reactions occur during the air-blowing process. The term "air-blown asphalt" is probably more correct.

One of the most significant properties of air-blown asphalt is that the temperature at which the asphalt becomes soft is raised. This temperature is called the softening point. Even at this temperature, asphalt still retains its waterproofing properties and durability. However, compared to paving-grade asphalt binders there is some reduction in ductility.

Air-blown asphalts are used for a wide variety of industrial and specialty purposes, including roofing applications, pipe-coating enamels, undersealing applications (to fill cavities beneath Portland cement concrete pavements), and waterproofing membranes for lining canals and reservoirs. In many instances, air-blown asphalts are applied in pure form. In other cases, very fine mineral powders (fillers) are mixed with air-blown asphalt before application.

▪ Roofing

▪ Properties, specification, and tests
Tests
Specifications

Roofing

One of the principal uses of air-blown asphalt is in roofing applications, whether used directly as mopping-grade asphalts in a built-up roofing (BUR) application or as saturants and coatings for rolled roofing felts and roofing shingles. As shown in Figure 1.6, approximately 10 percent of all asphalt (bitumen) used globally is used in roofing applications. In the United States, the use is slightly higher at approximately 15 percent.

Roofing asphalts and asphalt roofing materials (such as shingles) are exposed to a wide range of temperatures during application and in-service. During application, the temperature of the asphalt may be as high as 260°C (500°F). Depending on the geographical area in which it is used, an asphalt roofing material could experience temperatures as high as 80°C (176°F) and as low as –40°C (–40°F). Daily and seasonal temperature fluctuations produce thermal stresses and strains which roofing materials must tolerate without experiencing premature distress. When failures do occur, they may manifest in inadequate adhesion and cohesion, which can lead to flow, slippage, and cracking. These distresses can further lead to blistering, corrugating, splitting, and sliding.

In addition to temperature effects, roofing materials can experience an increase in aging due to photodegradation from ultraviolet (UV) light. The combined effects of temperature and UV exposure lead to aging of the asphalt materials which can affect their performance on the roof.

Properties, specification, and tests

The purpose of any effective materials specification is to provide tests and criteria that relate to the ultimate performance in-service. Although similar in many respects to the asphalt binder grades used for paving, the air blowing process provides materials that soften at higher temperatures than asphalt cements. Because the higher softening point is a most important and desirable property of air-blown asphalts, they are usually classified in terms of the ring-and-ball softening point test, rather than viscosity or penetration.

While blown asphalts are graded on the basis of the softening point, there are still penetration test requirements at three temperatures 32°F, 77°F, and 115°F (0°C, 25°C, and 46°C). These requirements provide a degree of control over the temperature susceptibility, or the rate of consistency change with temperature, for these materials. Applications for air-blown asphalt and associated specifications include the following:

- Mopping Asphalts—specified in ASTM D 312, these are used in BUR applications;
- Dampproofing and Waterproofing—specified in ASTM D 449;
- Canal, Ditch, and Pond Lining—specified in ASTM D 2521; and
- Undersealing—ASTM D 3141.

In addition to mopping grade asphalts used in roofing applications, saturants and coating grade asphalts are also used, although no national specification exists for either. Saturants are typically used for impregnating felts that are used in BUR as well as shingle and rolled products. Coating grade asphalts are used in the production of roofing shingles. Both coatings and saturants are similar to mopping grade asphalts, with saturants similar to, but softer, than the softest mopping grade asphalt (Type I); and coating grade asphalts similar to, but harder, than the hardest mopping grade asphalt (Type IV).

Tests

The specifications for industrial asphalts use primarily the same tests as the older specifications for paving grade asphalts. Despite efforts in the early 1980s to convert industrial application specifications from empirical tests (softening point, penetration) to more fundamental tests (viscosity), the specifications remained essentially unchanged.

Ring-and-Ball Softening Point

The Ring-and-Ball (R&B) Softening Point test is used as the basic measurement of consistency for grading air blown asphalts. In the test, described in ASTM Test Method D36, asphalt samples are poured into small brass rings (Figure 6.1); trimmed (Figure 6.2); and loaded with a

Figure 6.1 Filling R&B molds

steel ball in the center of each ring. After the samples are prepared, the assembly is suspended in a beaker of water, glycerin, or ethylene glycol at a height of 1 inch (25 millimeters) above a metal plate. The liquid is then heated at a rate of 5°C (9°F) per minute. As the asphalt softens, the balls and the asphalt gradually sink toward the plate. At the moment the asphalt touches the plate, the temperature of the liquid is determined, and this is designated as the ring-and-ball (R&B) softening point of the asphalt binder sample.

Figure 6.2 Trimming R&B specimens

Figure 6.3 SP apparatus (manual)

The R&B softening point test can either be performed manually or by using an automatic testing device. When using the manual method, the beaker is set on a ring stand with a Bunsen burner underneath it to control the rate of temperature increase (Figure 6.3). If using the automatic test equipment, the beaker is set on a heating element and heated at the prescribed rate (Figure 6.4).

Another softening point test that may be used with air-blown asphalts is the Mettler Cup-and-Ball Softening Point test. This test is similar in operation to the R&B Softening Point test and is used in the determination of the softening point of asphalt in the range from 50°C to 180°C. The Mettler Cup-and-Ball Softening Point test gives results comparable to those obtained by the R&B Softening Point test.

Figure 6.4 Ring and ball softening point (automatic)

It is important to note that for asphalt materials, the softening point is not the same thing as the melting point. Asphalts do not have an exact melting point since they are neither crystalline nor chemically homogeneous. As such, the softening point is simply an arbitrary indication of the consistency of the asphalt.

Penetration

For air-blown asphalts, the penetration test is conducted at three test temperatures as part of the specification. The test is the same procedure as that used for paving asphalt binders except that the total load and testing time varies. At the standard test temperature of 77°F (25°C), a total load of 100 grams is used for five seconds. At the higher test temperature of 115°F (46°C), a 50-gram load is used for five seconds. At the low test temperature of 32°F (0°C), a total load of 200 grams is used for 60 seconds.

Loss on heating

The loss on heating test is generally similar to the thin film oven test described in Chapter 4. The only differences are in the dimensions of the asphalt sample. Whereas the asphalt sample in the TFO test is about 5.5 inches (140 millimeters) in diameter and ⅛ inch (3 millimeters) deep, the sample for the loss on heating test is approximately 2.2 inches (55 millimeters) in diameter and 1.4 inches (35 millimeters) deep. In both tests, the asphalt and container are placed on a rotating shelf in a ventilated oven and maintained at 325°F (163°C) for 5 hours. The shelf rotates at approximately 5 to 6 revolutions per minute.

As with the TFO test, the loss on heating also conditions asphalt for other tests. It subjects the asphalt to hardening conditions simulative of the application processes. A penetration test usually is made on the air-blown asphalt binder sample after the loss on heating test for comparison with the penetration of the asphalt before the test.

Rotational viscosity

Although it is not specified, the rotational viscosity test is often conducted by manufacturers to determine the equiviscous temperature (EVT) for the application of roofing asphalt. The test is conducted in essentially the same manner as the test for paving asphalts. For mopping applications, the EVT is the temperature at which the apparent viscosity of the roofing asphalt is 125 centipoise (0.125 Pa-s). For mechanical spreader applications, the EVT is the temperature at which the apparent viscosity of the roofing asphalt is 75 centiPoise (0.075 Pa-s).

To determine the EVT of a roofing asphalt, it is necessary to have rotational viscosity measurements at a minimum of two temperatures. Depending on the type of roofing asphalt binder, temperatures may be as low as 163°C (325°F) and as high as 232°C (450°F). Once two measurements have been obtained, a plot is created showing viscosity as a function of temperature. For ease of graphing, viscosity is plotted on the y-axis by taking the logarithm of the viscosity (in centipoises) and then taking the logarithm again of that result. Temperature is plotted on the x-axis by taking the logarithm of temperature. The result is a double log-log graph that will result in a straight line between the data points.

Specifications

Specifications for mopping grade asphalts are given in ASTM D 312, *Standard Specification for Asphalt Used in Roofing*. According to the ASTM standard, the specification ". . . covers 4 types of asphalt intended for use in built-up roof construction, construction of some modified bitumen systems, construction of bituminous vapor retarder systems, and for adhering insulation boards used in various types of roof systems." As discussed earlier, asphalts are classified as Type I, Type II, Type III, and Type IV. These asphalts may also be referred to as dead level (Type I), flat (Type II), steep (Type III), and super-steep (Type IV) asphalts. Dead level roofing describes a roofing system in which there is a 0-2 percent incline. Low slope roofing describes a roofing system in which the incline of a roof is 14 degrees or less. Steep slope roofing describes a roofing system in which the incline of a roof is greater than 14 degrees.

Specifications for waterproofing applications are given in ASTM D 449, *Standard Specification for Asphalt Used in Dampproofing and Waterproofing*. This specification includes asphalt intended for use as a mopping coat in dampproofing, or for use in the construction of membrane waterproofing systems with felts and fabrics. In the specification, there are 3 types of asphalts. Type I readily flows under the mop and is suitable for use below grade in moderate temperature conditions. Type II is suitable for use above grade in warmer temperatures. Type III is the least temperature susceptible and is capable of being used above grade on vertical surfaces exposed to direct sunlight. Although the criteria are different, the tests are identical as those used in ASTM D 312.

Specifications for applications in canal, ditch, and pond lining are given in ASTM D 2521, *Standard Specification for Asphalt Used in Canal, Ditch, and Pond Lining*. This specification covers oxidized asphalt used as a waterproof membrane for canal, ditch, and pond linings. Tests are identical to the tests in ASTM D 449 with the addition of a requirement on loss on heating and retained penetration. The criteria in the specification are similar to a D 449 Type I asphalt for penetration and ductility and a D 449 Type III asphalt for softening point.

Finally, air-blown asphalt may also be used for undersealing Portland cement concrete pavements. Specifications are given in ASTM D 3141, *Specification for Asphalt for Undersealing Portland-Cement Concrete Pavements*.

Air-blown asphalt specification tests are shown in Table 6.1.

Table 6.1
Air-blown asphalt specification tests (ASTM)

	Test	D 312	D 449	D 2521	D 3141
Softening Point	D 36 or D 3461	x	x	x	x
Flash Point	D 92	x	x	x	x
Penetration	D 5				
0°C		x	x	x	x
25°C		x	x	x	x
46°C		x	x	x	x
Ductility	D 113				
25°C		x	x	x	x
Solubility	D 2042	x	x	x	x
Loss on Heating	D 6			x	
Retained Penetration @25°C (after loss on heating test)	D 5			x	

CHAPTER 7

Asphalt modification

The modification of neat asphalt binder to enhance its performance characteristics has occurred in the United States for more than 50 years. In a recent survey of 20 experts representing 18 states, 70 percent responded that there is a definite benefit in using polymer-modified asphalt mixtures to extend the pavement's service life. Nearly 60 percent of those experts also responded that the use of polymer-modified asphalt mixtures significantly reduced maintenance costs, but there has been insufficient performance data to quantify that benefit or enhancement.

The survey also indicated that the primary reason why users choose to use polymer-modified asphalt is to increase the mixture's resistance to rutting. Secondary reasons are to increase resistance to thermal cracking and to increase durability of the mixture.

It is important to note that while some user agencies have indicated a willingness to increase service life or reduce the risk of the development of early distresses through the use of modified asphalt binders (and their higher initial construction costs), not all asphalt pavements need to be constructed using modified asphalt mixtures. Each project should be evaluated to determine if the environmental conditions, traffic loading, expected service life, and performance warrant the use of modified asphalt materials.

- **Types of modifiers and additives**
- **Additional tests for modified asphalt binders**

Types of modifiers and additives

Modified asphalt binders are sometimes generically referred to as "polymer-modified asphalts" or PMAs. In actuality, modified asphalt binders may be produced in a number of ways including polymer and chemical modification, although polymer modification is the most prevalent.

Polymers are made up of smaller molecules, or monomers, that are chemically connected to form long molecular chains. Names of polymers are usually based on the names of the monomers used to make the specific polymer. Thus, the polymer that comes from the polymerization of ethylene monomers is called "polyethylene."

Most often, when monomers are polymerized, they form long-chain polymers. Some types of these polymers, called thermoplastics, soften when heated, enabling them to be re-formed. Some examples of thermoplastic polymers include nylon, polyethylene, and polyethylene terephthalate (which is used in clear plastic bottles).

By contrast, some monomers, when polymerized, connect to form a three-dimensional network. These types of polymers, called thermosets, are permanently (irreversibly) set in shape. Some examples of thermoset polymers include epoxy resins and vulcanized rubber.

Two broad classes of polymers used in asphalt modification are polyolefins and styrenic polymers. Polyolefins are so named because they are based on the polymerization of molecules containing a simple double bond or olefin. Polyolefin polymers include polyethylene, polypropylene, and ethylene vinyl acetate. Styrenic polymers are so named because they are based on polystyrene that has been copolymerized with other small molecules—most commonly butadiene.

Polymers can also be classified based on physical properties. Depending on their behavior when stretched with sufficient force, polymers are classified as plastomers (plastics) or elastomers (rubbers or elastics). When stretched, plastomers will yield and remain in their stretched position when the load is released. Elastomers will yield under load (stretch) but will return to their original shape when the load is released. Generally speaking, most polyolefins behave as plastomers, while styrene-butadiene copolymers behave as elastomers.

Finally, when blended into asphalt, polymers can behave in two different ways. If the polymer forms discrete particles in the asphalt binder, then it functions primarily as a thickener (filler), which increases the viscosity of the asphalt binder while having no significant effect on low temperature properties. If the polymer forms a continuous network in the asphalt binder, then it functions as a homogeneous blend, which may impart some of the physical characteristics of the polymer to the asphalt binder. In this instance, the high- and low-temperature properties of the asphalt binder may be affected.

Elastic modification (elastomers)

The most common comonomer used to produce styrenic polymers is butadiene. When styrene and butadiene are polymerized in a random arrangement, referred to as a "random copolymer," the polymer is called styrene-butadiene rubber (SBR). This type of polymer may also be referred to as synthetic latex. Figure 7.1 provides a microscopic view of SBR modification of asphalt.

When styrene and butadiene are polymerized in discrete, connected blocks, referred to as a "block copolymer," the polymer is called styrene-butadiene (SB) or styrene-butadiene-styrene (SBS). Figure 7.2 shows SBS in the form of pellets.

Although some elastomers are blended inline at the hot mix asphalt mixing facility, the most common practice is to pre-blend the polymer and asphalt binder at the terminal or refinery (see Figure 7.3). Some PMA systems require only mixing of the polymer into the asphalt binder in a mixing kettle, while others require high shear milling or other special mixing operations. These processes take place at specific temperatures and are continued until the materials are appropriately mixed.

Surveys conducted during the NCHRP 9-10 research (entitled "Superpave Protocols for Modified Asphalt Binders") in 1996 indicated that elastomers, and more specifically SBS polymers, were the most commonly used modifiers by user agencies. In their responses, most users indicated

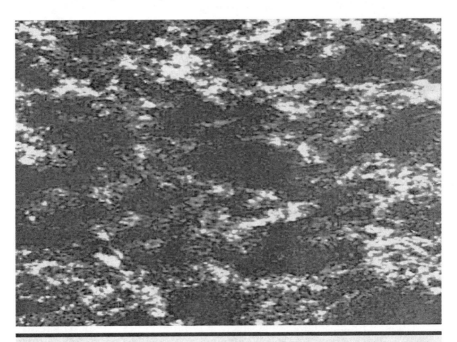

Figure 7.1 SBR Latex Modification of Hot Asphalt Cement

Figure 7.2 SBS Pellets

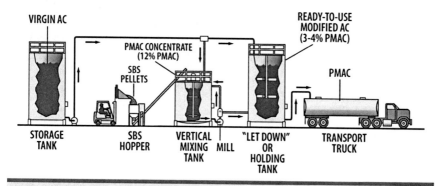

Figure 7.3 SBS blending system for Asphalt Terminals

that they selected elastomeric modifiers to help mitigate permanent deformation or rutting of asphalt pavements. Mitigation of fatigue and low-temperature cracking was also seen as a potential benefit.

In the past, specifications in the United States have sought to characterize the elastic properties of a styrenic polymer-modified asphalt binder through the use of tests that stretch the asphalt binder and measure the stress-strain response (as in the Force Ductility test and the Toughness and Tenacity test) or the recovery (as in the Elastic Recovery test). The

Softening Point test has often been used to evaluate high-temperature resistance to deformation. As the percentage of styrenic polymer increases in the asphalt binder, it is expected that the high-temperature stiffness will increase as will the stress-strain and recovery response.

Plastic modification (plastomers)

Plastomers are a class of polymers used in asphalt binders that, when stretched, will yield and remain in their stretched position when the load is released. These polyolefin polymers add to the high-temperature stiffness of the asphalt binder but will not have the elastic characteristics of styrenic polymers. Polyolefin polymers that are commonly used to modify asphalt binders include polyethylene (often the low density polyethylene identified as LDPE) and ethylene vinyl acetate (EVA). In this class of polymers, EVA has been most commonly used in the United States. As with elastomeric modifiers, most users indicated in the survey that they selected plastomer modifiers to help mitigate permanent deformation or rutting of asphalt pavements.

Crumb rubber modification

Although it is an elastic material, crumb rubber—produced from ground tires—is considered differently. Crumb rubber is incorporated in an asphalt mixture either as an aggregate modifier (the "dry process") or as a modifier to the asphalt binder. Since the "dry process" is a mixture modifier rather than an asphalt modifier, it will not be further addressed in this section.

Two principal methods exist for incorporating crumb rubber as a modifier (CRM) to the asphalt binder. The "wet process" (also referred to as the "McDonald process" named for the first patented technology that applied the wet process) is conducted by pre-blending and reacting crumb rubber of a relatively small size (usually 600 microns or smaller) in the asphalt binder at an elevated temperature for a specific time to allow for reaction. The wet process, also referred to as Asphalt-Rubber, is defined in ASTM as follows:

"A blend of asphalt cement, reclaimed tire rubber, and certain additives in which the rubber content is at least 15 percent by weight of the total blend and has reacted in the hot asphalt cement sufficiently to cause swelling in the rubber particles."

The size of the CRM particles may vary, but it is a generally recommended practice that all CRM particles be finer than the 2.36-millimeter (#8) sieve. The wet process can be used to make CRM asphalt binders with a range of physical properties, with the most important distinction being the rotational viscosity of the blend at high temperature (either 177°C or 190°C, depending on the applicable specification). Studies have shown that viscosity in a CRM asphalt binder is strongly related to the CRM particle size, reaction time, and tire rubber content in the blend. CRM asphalt binders that have a rotational viscosity in excess of 1.5 Pascal-seconds at the specified high temperature should be assumed to require agitation to keep the CRM particles uniformly dispersed. For this reason, many CRM asphalt binders produced using the wet process are produced at the asphalt mixing plant where the CRM asphalt binder can be continuously agitated until use.

The Terminal Blend Process produces CRM asphalt binders in the same general category as the wet process, with the difference being that Terminal Blend CRM asphalt binders are produced using much smaller CRM particle sizes in a process that facilitates digestion of the CRM into the asphalt. As such, these asphalt binders do not require constant agitation to keep discrete rubber particles uniformly distributed. This means that the CRM asphalt binder can be blended at the terminal and shipped to the asphalt mixing plant or project site without concerns about product homogeneity. It also means that the asphalt binder can be tested and graded using the PG equipment and procedures required in AASHTO M 320. Terminal Blend CRM asphalt binders are typically modified with fine CRM particles passing the 0.3-millimeter (#50) sieve that can be digested relatively quickly and kept dispersed using normal circulation. Typically, these Terminal Blend CRM asphalt binders do not have a viscosity that exceeds 1.5 Pascal-seconds at high temperature.

CRM asphalt binders have been used in a variety of applications. In addition to paving grade asphalt binders used in an asphalt mixing plant, they are used in such applications as chip seals, joint sealers, and stress-absorbing membrane interlayers (SAMI). Research has shown that the properties of Wet Process CRM asphalt binders depend on:

- Rubber source and processing method (ambient or cryogenic)
- CRM particle size
- CRM concentration
- Asphalt binder source and grade
- Asphalt additive type and concentration
- Interaction time and temperature

In the United States, crumb-rubber modification is used extensively by some users and not at all by other users. Generally, crumb-rubber modification is considered to address the same distresses (permanent deformation) as elastomeric modification.

Mixed modification

Mixed-polymer modification, such as the combination of styrenic and polyolefin polymers, is not commonly used. Potential chemical incompatibility and the cost of using two polymers in the asphalt modification process may make the combination undesirable. However, it is possible that the addition of an elastomeric polymer to a plastomeric modifier could enhance low-temperature characteristics while maintaining high-temperature stiffness.

Chemical modification

Chemical modification refers to any of a number of chemicals that might be added to an asphalt binder to change its properties. Most often, chemical modification involves the addition of polyphosphoric acid (PPA) to an asphalt binder, either alone or in combination with other types of polymer modification.

PPA is a liquid mineral polymer that may be produced from orthophosphoric acid by removing water through condensation. In removing the water, the PPA has a higher density and viscosity, blends well with asphalt, and has significantly lower corrosivity for steel and stainless steel compared to orthophosphoric acid.

In general, polyphosphoric acid increases the high-temperature stiffness of an asphalt binder. However, the effect depends on the concentration and the source (properties) of the crude used to produce the asphalt binder.

Extenders

Asphalt binder extenders include materials such as sulfur and wood lignin. Of the two, sulfur has been the most extensively used and evaluated in asphalt mixtures in the United States—primarily from 1973 to 1985. At that time, asphalt binder and sulfur prices created a favorable situation, enabling users to consider substitution of asphalt binder with sulfur. Although laboratory and field studies indicated a general increase in the high-temperature stiffness of asphalt mixtures incorporating sulfur modification, economic considerations eventually resulted in its reduced use. In recent years, however, there has been an increased interest once again in the use of sulfur modification of asphalt binders and mixtures. Sulfur may also be used in combination with certain elastomeric polymers, such as SBS, to create cross-linking in the asphalt binder. This cross-linking results in a chemical bonding between the polymer molecules, creating a continuous network in the asphalt binder which leads to a change in physical properties (usually a greater elastic response).

Oxidants and antioxidants

Oxidants principally include manganese compounds and are used to artificially "oxidize" the asphalt binder, thereby increasing its stiffness. These types of materials were used experimentally in the United States in the 1930s and later in the 1980s, but have not been commonly used since. While the initial increase in stiffness may be beneficial from the standpoint of a permanent deformation resistance in an asphalt mixture, the early increase in stiffness means that the asphalt will act as if it were prematurely aged—meaning cracking could show up earlier than if the oxidant had not been used as a modifier.

Antioxidants serve the opposite purpose, in that they are intended to minimize the oxidative hardening of asphalt binders. Most typically, these types of materials include carbamates (lead, zinc), carbon black, and hydrated lime. Of this class of materials, hydrated lime has been the most commonly-used modifier, although most users select hydrated lime for its benefits in reducing the moisture susceptibility of asphalt mixtures irrespective of any antioxidant benefits it may have.

Hydrocarbons

Hydrocarbon modifiers generally fall into two generic categories: hardening modifiers and softening (or rejuvenating) modifiers. In the former category are natural asphalts and refined asphalts. Natural asphalts that have been used as modifiers include Gilsonite (a zero-penetration resin) and Trinidad Lake Asphalt (TLA). Gilsonite is added as a dry additive to either the asphalt binder or mixture. TLA is slightly softer than Gilsonite and is similar to some industrial, air-blown asphalts, except that it contains a significant amount of inert organic matter. Refined asphalts may include solvent de-asphalted bottoms such as ROSE unit asphaltenes. In either case, hydrocarbon hardening modifiers are used to increase the high-temperature stiffness (resistance to permanent deformation) of the asphalt binder.

Softening or recycling modifiers include naphthenic or aromatic oils. These modifiers may be added to the asphalt binder or to the asphalt mixture separately from the asphalt binder. Most often, they are used in asphalt mixtures containing reclaimed asphalt pavement (RAP) material as they serve to lower the viscosity of the aged asphalt binders.

Antistripping additives

Antistripping additives are used to reduce or eliminate the moisture susceptibility of an asphalt mixture by promoting adhesion between the asphalt binder and aggregate surface. Liquid antistripping additives principally include amines such as polyamines and fatty amidoamines. Most often, the practice is to blend the liquid antistripping additive with the asphalt binder at the refinery or terminal, but in any case before mixing with aggregate. The antistripping concentration is typically 0.5 percent of the total weight of the asphalt binder. A slight decrease in viscosity may occur with the addition of the additive (which is fluid at room temperature).

Additional tests for modified asphalt binders

In addition to conventional specification tests common for all asphalt binders, modified asphalt binders often have additional tests that are required to verify the presence of polymer, and/or to characterize their properties. These range from separation tests that characterize the compatibility of the asphalt-polymer blend to recovery and stress-strain tests that characterize the elastic (or delayed-elastic) behavior of the modified asphalt binder.

Separation tests—ointment tube

When a polymer-modified asphalt is blended and stored in a tank at elevated temperatures for a period of time, some thermal degradation may occur. A commonly used test to assess polymer-asphalt compatibility is generally referred to as the "ointment" or "cigar" tube test. In this test, an aluminum ointment tube of a specific size is filled with 50 grams of modified asphalt binder (Figure 7.4). The filled tubes are then sealed and allowed to stand vertically in an oven operating at 163°C for 48 hours. The tubes are then transferred to a freezer for 4 hours. After freezing, the tubes are removed and cut into thirds. The top third and bottom third are separated

Figure 7.4 Ointment tubes

into tins and tested using either the softening point or DSR tests to determine if a significant difference exists in the properties of the top and bottom thirds. A significant difference in test values indicates that the polymer may have separated from the asphalt binder or have thermally degraded.

Solubility—centrifuge

The solubility test described earlier was first introduced into a specification in 1903. The solubility test for petroleum asphalt binders, as it exists today in asphalt binder specifications, is used to ensure that an asphalt binder does not contain insoluble organic and inorganic matter that doesn't contribute to the "active cementing constituents" of the asphalt binder. As the use of polymer-modified asphalt binders has increased, a different solubility test (ASTM D 5546) was developed. ASTM D 5546 is similar to the standard solubility test except that it uses a centrifuge and toluene as the solvent. It was developed principally because some polymer modifiers did not dissolve adequately during the standard solubility test procedure, thus resulting in some polymer-modified asphalt binders failing the solubility criterion.

Recovery and stress-strain tests

To evaluate whether an asphalt binder has been modified with an elastomer—specifically a block copolymer such as SBS—specifying agencies will often add tests designed to measure the ability of the asphalt binder to stretch when loaded and ultimately rebound to its original shape when unloaded. The ability, and extent, to which an asphalt binder will stretch and recover is related to the level of elastomeric modification in the asphalt binder.

Three tests are commonly used in the United States for testing elastomeric-modified asphalt binders: elastic recovery, force ductility, and toughness and tenacity. Each test involves stretching an asphalt binder sample and measuring either the stress-strain response or recovery amount. An additional parameter used by some states is the phase angle measured using the dynamic shear rheometer. The phase angle is an indicator of the relative amount of the elastic component in the asphalt binder.

An additional test proposed by the NCHRP 9–10 research for characterizing modified asphalt binders was the DSR Repeated Creep Test. This test has been modified as the multiple stress creep recovery (MSCR) test.

Elastic Recovery

The Elastic Recovery (ER) test, as detailed in AASHTO T 301, is a simple procedure based on the ductility test equipment. In the test, a specimen is prepared as in the ductility procedure, except that the side pieces are flat, rather than V-shaped (Figure 7.5).

To conduct the test, a heated sample of asphalt binder (usually RTFO-aged) is poured into the mold, slightly overfilling the mold. After a prescribed cooling period, the specimen is trimmed flush with the surface of the mold. The trimmed test specimen is placed in the ductility water bath and conditioned to the desired test temperature, usually 77°F (25°C). After the conditioning period, the specimen is loaded into the ductility machine and one end of the specimen is pulled away from the other at a rate of 5 centimeters per minute until the specimen has stretched a prescribed length—either 10 or 20 centimeters (Figure 7.6). Depending on the procedure, the specimen is then either immediately cut in the center or held in a stretched position for five minutes before cutting in the center (Figure 7.7). After one hour, the specimen is pushed back together until the cut ends touch (Figure 7.8). This final measurement is then used to calculate the recovery of the asphalt binder specimen. Higher recovery values are indicative of an asphalt binder with a greater delayed elastic response.

Force Ductility

Like the Elastic Recovery test, the Force Ductility test (as detailed in AASHTO T 300) is a simple procedure based on the ductility test equipment. In the test, a specimen is prepared as in the ductility procedure, except that the side pieces are flat, rather than V-shaped.

The specimen preparation and test procedure is essentially identical to the Elastic Recovery test. The main differences between the tests are the following:
- A load cell is used to record the force required to pull the asphalt binder specimen.

Figure 7.5 ER test specimen

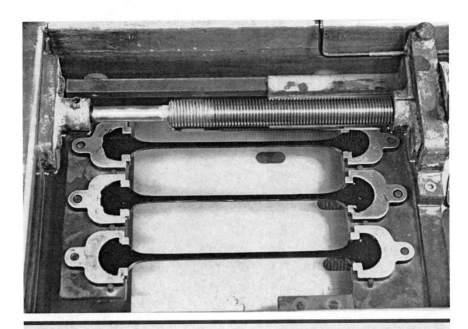

Figure 7.6 ER test operation

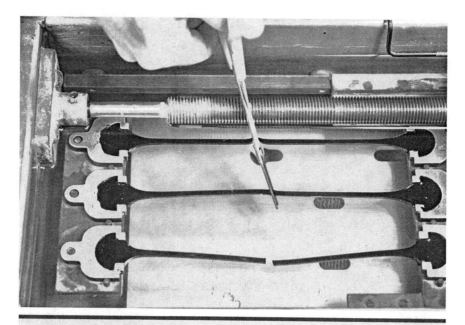

Figure 7.7 ER specimens cut

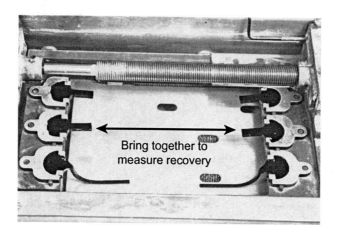

Figure 7.8 ER specimens recover

- The test is continued to 30 centimeters (Figure 7.9), at which point the test ends.
- Test temperature is usually 41°F (4°C).
- The asphalt binder is usually tested in the original (unaged) condition.

In the test, a stress-strain curve is constructed (Figure 7.10). Typically, the stress at 30 centimeters is divided by the peak stress to generate a force ductility ratio. Specimens that do not stretch to 30 centimeters before

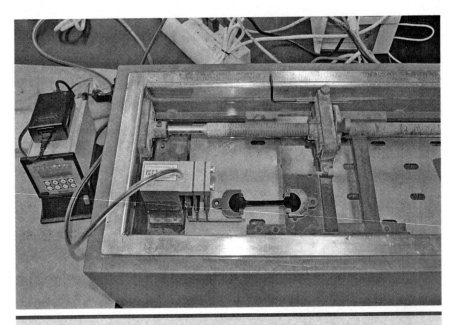

Figure 7.9 FD test operation

breaking are considered to fail the test. Increasing force ductility ratios are indicative of increasing elasticity in the asphalt binder.

Toughness and tenacity

A third test conducted by some agencies in the United States is the Toughness and Tenacity test. This procedure was first introduced by Benson in the 1950s. In the test, a metal hemispherical head is embedded in hot asphalt to a depth of approximately 11 millimeters. After cooling to 77°F (25°C), the head is attached to a tensile test machine and pulled from the asphalt binder at a rate of 51 centimeters per minute (Figure 7.11). The load is measured throughout the test and a load-deformation curve is plotted (Figure 7.12). Toughness and tenacity values are determined based on the area under different portions of the load-deformation curve.

DSR phase angle

In the standard DSR test used for performance-graded (PG) asphalt binders, the complex shear modulus (G^*) and phase angle (δ) are determined. The phase angle, δ, is a measure of the elastic and viscous components of the asphalt binder. Some user agencies specify a maximum phase angle for modified asphalt binders. Unfortunately, the phase angle is influenced by the stiffness of the asphalt binder. Thus, it is not a true reflection of the elasticity—in the form of an elastomer—of the asphalt binder.

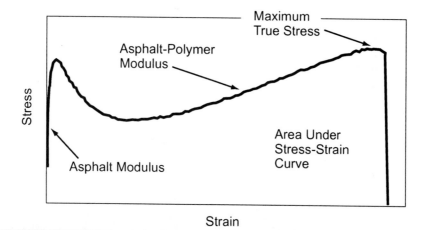

Figure 7.10 FD output

Figure 7.11 T&T equipment and operation

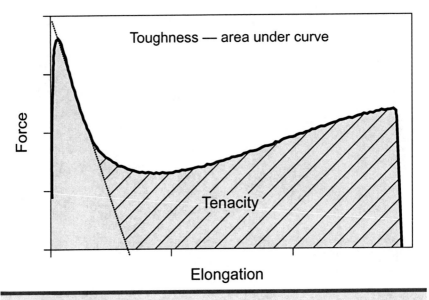

Figure 7.12 T&T output

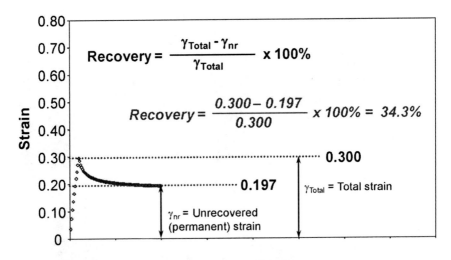

Figure 7.13 Calculating MSCR recovery

DSR Creep-Recovery

Recently researchers have evaluated the properties of modified asphalt binders using the DSR in a repeated creep-recovery test. This test has been modified to include multiple stress levels. It is now referred to as the multiple-stress creep-recovery (MSCR) test. It can be used to determine

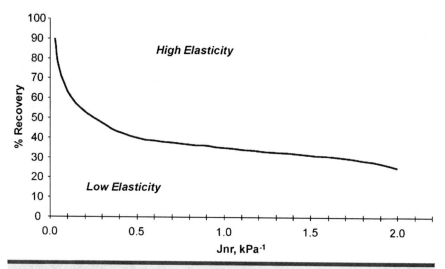

Figure 7.14 Determining elastic behavior from MSCR

the amount of recovery in an asphalt binder during creep-recovery testing—in addition to being used to determine the nonrecoverable creep compliance, J_{nr}.

The MSCR test procedure is described earlier. After the average amount of recovery is calculated (Figure 7.13), the results are used in combination with J_{nr} to indicate whether an asphalt binder has a significant elastic component (Figure 7.14).

It is important to know both the J_{nr} and Recovery of the modified asphalt binder to know if it has significant elastic properties.

CHAPTER 8

Safety

The information in this chapter is provided to give the reader a *very* cursory overview of safety concerns related to the handling of asphalt binders. For more detailed information, please see other Asphalt Institute sources such as *IS-230, The Bitumen Industry–A Global Perspective,* and *VA-26, Safe Handling of Hot Asphalt.*

▪ Safe handling of asphalt
    Storage temperatures
    Asphalt fume

Safe handling of asphalt

Because asphalt products are often stored and handled at elevated temperatures, fire prevention is extremely important. One of the greatest hazards in handling hot asphalt is exposure to a source of ignition. Sources of ignition, such as sparks of electrical or other origin, open flames, or incandescent material (lighted cigarette) should be prohibited or otherwise strictly controlled in the vicinity of asphalt operations.

Many of the metal surfaces in asphalt plants and in roofing or processing operations exceed 150°F (65°C) and often are well over 200°F (93°C). Asphalt temperatures commonly exceed 300°F (149°C) as used in these areas. Exposed lines or surfaces can burn flesh on momentary contact at 150°F (65°C) or higher. Also, similar hazards exist in other areas: around dryers, boiler houses, asphalt receiving and handling areas, and similar locations. It is very important that workers be aware of burn hazards in the workplace. Burn hazards should be clearly indicated by warning signs. Avoid situations where hot asphalt can be spilled or splashed. Always wear appropriate protective clothing and follow proper safety procedures. Use catwalks, screens, barrier guards, and shields to protect from steam, hot asphalt, hot surfaces, and similar dangers.

Never enter an empty asphalt tank unless it is properly ventilated and has been purged of asphalt fumes. It is advisable to have a second person in the immediate area as an observer in case one loses consciousness due to lack of oxygen.

Whenever a person is injured from exposure to asphalt fumes or hot asphalt, obtain first aid/medical attention immediately. If hot asphalt comes in contact with the skin, the affected area should be immediately immersed in cool water to cool the asphalt and stop the burning of the skin. Do not attempt to wipe off the hot asphalt with a cloth because this may remove skin along with the asphalt. To prevent the possibility of future medical complications, have the victim examined by a physician even if the injury does not appear to be serious.

Storage temperatures

From a safety standpoint, it is desirable to store asphalt at a temperature as far as possible below the flash point. However, it should be recognized that the values reported by the flash point tests are specific to the test procedures employed and not necessarily representative of the vapor space atmospheres existent in storage.

Storage temperatures in the 212°F–266°F (100°C–130°C) range are particularly troublesome. Storage temperature should either be below 212°F (100°C) or above 266°F (130°C). In particular, temperatures should not be allowed to cycle above and below the boiling point of water (approximately 212°F or 100°C, depending on elevation). See Table 10.1 in Chapter 10 for typical storage and mixing temperatures for paving grade asphalt binders.

Asphalt fume

During the handling of asphalt at elevated temperatures, small quantities of hydrocarbon emissions are released. Asphalt fume is the material which is measured and reported as representing potential occupational exposure. Asphalt fume is a complex mixture having a broad range of boiling points. The molecular composition can include hydrocarbons (from naphtha to long-branched chain aliphatic hydrocarbons), cycloalkanes, aromatics, and hetero-molecules containing sulfur, oxygen, nitrogen, and low levels of polynuclear aromatic compounds (PAC), including polynuclear aromatic hydrocarbons (PAH). In a lab study of paving grade asphalt binders it was found that, within the relevant temperature range (140–190°C), the emission rate of the Benzene Soluble Fraction doubled for every 11–12.5°C increase in temperature. Similarly for roofing asphalt binders, within the relevant temperature range (210–270°C), the emission rate of the Benzene Soluble Matter (BSM) doubled for every 14–17°C increase in temperature. Thus, exerting proper temperature control when heating asphalt binders can, consequently, control the emission rate of BSM.

CHAPTER 9

Understanding testing variability

Variability is defined as the standard error that occurs in any process that makes one sample different to some extent from another sample. It can include inherent material variability, sampling variability, and testing variability. As technologists involved in testing, we can control sampling variability by ensuring that we have exercised good sampling procedures. We can control testing variability by ensuring that our equipment is properly calibrated and operational and that we closely follow standardized test procedures and good laboratory practices. As testing technologists, the only thing we cannot control is inherent material variability.

- **Repeatability and reproducibility**
    Accuracy, precision, and bias
- **Variability in asphalt binder tests**
- **Dispute resolution**

Repeatability and reproducibility

The American Society for Testing and Materials (ASTM) defines a test method as "... a definitive procedure for the identification, measurement, and evaluation of one or more qualities, characteristics, or properties of a material ... that produces a test result." For any test method, we want to know how closely test results compare when repeated tests are conducted on the same material. Within-laboratory precision, which is often referred to as repeatability, is a comparison of test results conducted at separate times on the same material in the same lab on the same equipment by the same technician. Between-laboratory precision, which is often referred to as reproducibility, is a comparison of test results conducted at separate times on the same material in a different lab on a different set of equipment and by a different technician.

For testing laboratories, technologists are concerned not only about repeatability within the lab but also reproducibility with other labs. In this instance, a statistical term called the "difference two-sigma limit," or d2s for short, is used to provide an estimate of the acceptable range of test results from two labs that can be expected as a result of normal testing variability. It represents the situation where the difference in results would be equaled or exceeded in 1 of 20 cases (5 percent) of normal, proper operation. It is determined by multiplying the calculated standard deviation (σ) by $2\sqrt{2}$, or 2.83. In cases where the standard deviation is proportional to the average for different levels of the measured property, the d2s percent is calculated as the d2s divided by the average of the 2 test results.

The d2s and d2s percent limits are useful in evaluating test results because it lets the technologist know how to respond to a set of test data in his/her lab compared to another lab. If the difference between the test results in Lab A and Lab B are within the d2s limits, then both results could be considered "correct" and the "actual" value could be somewhere between the 2 results. If, however, the difference between the test results in Lab A and Lab B exceeds the d2s limits, then a retest and possibly reevaluation of test equipment and procedures may be needed. Needless to say, the smaller the d2s or d2s percent, the more reproducible the test result.

Accuracy, precision, and bias

In discussing repeatability and reproducibility, it is also useful to define three terms that are related: accuracy, precision, and bias. Although they may be similar, accuracy and precision are most definitely not the same. Accuracy is a measurement of how close results come to the "actual" value. Precision is a measurement of how close replicate results are to each other.

To illustrate the concepts of accuracy and precision, let's use the example of throwing darts at a dart board. If a person is aiming for the bull's-eye and he hits it, then we say that the throw was very accurate. If the next 2 throws also hit the bull's-eye, then the set of 3 darts were not only accurate but also precise (they have good precision). It is also possible for all 3 darts to be grouped together tightly 3 inches from the bull's-eye. In this case, there still is good precision, but not good accuracy.

Bias is a special case where there is a consistent trend in test results as a function of the material or test. In the dart board example, bias would be exemplified by consistently missing the bull's-eye by 3 inches to the right on every throw. A "recalibration" of the release point would be needed to move the darts towards the bull's-eye.

Variability in asphalt binder tests

Most standardized test procedures now routinely require a statement about the repeatability (single-operator precision) and reproducibility (multi-laboratory precision) of the test results derived from the standard procedure. Generally speaking, any new test procedure should undergo ruggedness testing before attempting to determine repeatability and reproducibility.

When a test method is still early in its development, ruggedness testing evaluates whether minor variations in test parameters cause major variations in test results. As an example, suppose a test procedure says to hold the sample at temperature for 60 ± 5 minutes before testing. In a ruggedness evaluation, the sample would be tested at 55 and 65 minutes and the results compared to see if they are significantly different. If so, the tolerance of ± 5 minutes might need to be further tightened. According to ASTM C 1067, "... ruggedness testing has as its purpose the detection and control of sources of testing variation prior to programming an interlaboratory study. One of the most productive uses of a ruggedness or screening evaluation is the elimination of those test methods shown to have poor precision even after making vigorous efforts to reduce the variation."

After ruggedness testing is complete and the test procedure is determined to be rigorous, then an interlaboratory, or round-robin, study is usually conducted to determine repeatability and reproducibility. This interlaboratory study should include several different labs and materials with different properties representative of the range of materials that are expected to be tested. An example is shown in Table 9.1 for AASHTO T 313 (Bending Beam Rheometer).

Table 9.1
Precision statement for AASHTO T 313 – BBR

Test/Result	Single Operator Precision (Repeatability)		Multi-Laboratory Precision (Reproducibility)	
	1s %	d2s %	1s %	d2s %
BBR				
S(60)	2.5 %	7.2 %	6.3 %	17.8 %
m(60)	1.0 %	2.9 %	2.4 %	6.8 %

Note: The precision estimates are based on the analysis of eight pairs of AMRL proficiency samples. The data analyzed consisted of results from 174 to 196 laboratories for each of the eight pairs of samples. The analysis included five binder grades: PG 52-34, PG 64-16, PG 64-22, PG 70-22 and PG 76-22 (SBS modified). Average creep stiffness results ranged from 125.4 MPa to 236.8 MPa. Average slope results ranged from an m-value of 0.308 to 0.374. The details of this analysis are in NCHRP Final Report, NCHRP Project No. 9-26, Phase 3.

In Table 9.1, note that the variability introduced in a multi-laboratory comparison is always higher than the variability experienced by a single operator. Thus, when comparing data from 2 labs, the acceptable range of 2 results on BBR Stiffness, or S(60), is 17.8 percent. Within one lab with a single operator, the acceptable range of 2 test results on BBR S(60), is much lower (7.2 percent). To determine if 2 test results are within acceptable limits of variability, the technician simply needs to compare the 2 test results to the published d2s or d2s percent. An example is shown below using the data in Table 9.1.

Example 1

Two independent labs perform AASHTO T 313 testing on the same PG 64-22 asphalt binder sample at −12°C with results shown below:

	Lab A	Lab B
S(60), MPa	216	188
m(60)	0.313	0.321

Are the results within acceptable limits of variability? To determine this we need to follow these steps:

Step 1	*Determine the average of the 2 results for S(60):* $$\text{Average} = \frac{216 + 188}{2} = 202 \text{ MPa}$$
Step 2	*Multiply the average value of S(60) by the appropriate d2s percent (from Table 9.1, the multi-laboratory value for d2s percent is 17.8 percent):* $$\text{d2s} = 202 \times 17.8 \text{ percent} = 36.0 \text{ MPa}$$
Step 3	*Determine the difference between the 2 S(60) test results:* $$\text{Difference} = 216 - 188 = 28 \text{ MPa}$$
Step 4	*Compare the difference in the test results to the calculated d2s value. If the difference is less than or equal to the d2s value then the results are within acceptable limits of variability:* Difference = 28 MPa d2s = 36 MPa Since the difference between the test results is less than the calculated d2s value, the results are considered to be within acceptable limits of variability for S(60).

Repeating the calculations for m(60)...

Step 1	*Determine the average of the 2 results for m(60):* $$\text{Average} = \frac{0.313 + 0.321}{2} = 0.317$$

Step 2	Multiply the average value of m(60) by the appropriate d2s percent (from Table 1, the multi-laboratory value for d2s percent is 6.8 percent): d2s = 0.317 × 6.8 percent = 0.021
Step 3	Determine the difference between the 2 m(60) test results: Difference = 0.321 − 0.313 = 0.008
Step 4	Compare the difference in the test results to the calculated d2s value. If the difference is less than or equal to the d2s value then the results are within acceptable limits of variability: Difference = 0.008 d2s = 0.021 Since the difference between the test results is less than the calculated d2s value, the results are considered to be within acceptable limits of variability for m(60).

If d2s is provided instead of d2s percent, simply skip Steps 1 and 2 above.

As shown in the example, the variability of the two sets of data generated by different labs is within acceptable limits. However, just because the data is considered acceptable does not mean that there is no room for improvement in laboratory practices to improve reproducibility. If similar results are produced consistently on other samples (Lab A always having slightly higher stiffness and lower *m-value* than Lab B), then the two labs should work together to resolve possible sources of variability or bias in the lab testing process.

Table 9.1 also includes a footnote that describes the range of test values for which the precision analysis was conducted. For the BBR, the range of S(60) values was from 125 to 237 MPa and the range of m(60) values was from 0.308 to 0.374. The precision estimates are not necessarily invalid if material values are outside of these ranges (such as evaluating an asphalt binder sample with a stiffness of 425 MPa and an *m-value* of 0.255), but they should be used with caution and an understanding that the actual variability may be different.

In some cases, the published precision is not a single number but rather is dependent upon the value of the test result. This can either be shown by listing a range of values and the corresponding precision, as is done with the Penetration test (AASHTO T 49, ASTM D 5), or by describing an equation that relates the precision with the test result, as is done for the RTFO test (AASHTO T 240, ASTM D 2872).

For reference, Tables 9.2 and 9.3 contain current test variability information for several common asphalt binder tests. It is not an exhaustive list and should be used with some caution as the published variability may vary as further interlaboratory studies and analyses are completed. Always consult the latest versions of AASHTO and ASTM standards for the most current information and to see the conditions under which the precision statement was developed.

Table 9.2
Asphalt binder test variability (1s percent and d2s percent)

Test/Result	Test Method	Single Operator Precision (Repeatability)		Multi-Laboratory Precision (Reproducibility)	
		1s %	d2s %	1s %	d2s %
BBR	AASHTO T 313	S(60) 2.5 %	S(60) 7.2 %	S(60) 6.3 %	S(60) 17.8 %
		m(60) 1.0 %	m(60) 2.9 %	m(60) 2.4 %	m(60) 6.8 %
DSR	AASHTO T 315	G*/sin δ (Original) 2.3 %	G*/sin δ (Original) 6.4 %	G*/sin δ (Original) 6.0 %	G*/sin δ (Original) 17.0 %
		G*/sin δ (RTFO) 3.2 %	G*/sin δ (RTFO) 9.0 %	G*/sin δ (RTFO) 7.8 %	G*/sin δ (RTFO) 22.2 %
		G*sin δ (PAV) 4.9 %	G*sin δ (PAV) 13.8 %	G*sin δ (PAV) 14.2 %	G*sin δ (PAV) 40.2 %
Rotational Viscosity	AASHTO T 316	1.2 %	3.5 %	4.3 %	12.1 %
Direct Tension Fail. Stress Fail. Strain	AASHTO T 314	7.4 % 11.4 %	20.8 % 32.2 %	18.6 % 31.5 %	52.5 % 89.1 %
Flash Point (COC)	AASHTO T 48	3 %	8 %	10 %	28 %
TFO Mass Loss	AASHTO T 179	x > 0.4 % Loss: 2.9 %	x > 0.4 % Loss: 8 %	x > 0.4 % Loss: 14 %	x > 0.4 % Loss: 40 %
Absolute Viscosity	ASTM D 2171		7 %		10 %
Kinematic Viscosity	ASTM D 2170	0.64 %[a]	1.8 %[a]	3.1 %[a]	8.8 %[a]
Toughness and Tenacity	ASTM D 5801	Toughness 6.8 %	Toughness 20 %	Toughness 11.3 %	Toughness 32 %
		Tenacity 7.3 %	Tenacity 20 %	Tenacity 11.5 %	Tenacity 32 %
Saybolt-Furol Viscosity	AASHTO T 59		25°C 5 %		25°C 15 %
			50°C 9.6 %		50°C 21 %

[a] Results for asphalt binders tested at 135°C. Please refer to method for precision statement for asphalt materials tested at 60°C.

Table 9.3
Asphalt binder test variability (1s and d2s)

Test/Result	Test Method	Single Operator Precision (Repeatability)		Multi-Laboratory Precision (Reproducibility)	
		1s	d2s	1s	d2s
RTFO Mass Loss	AASHTO T 240	0.0061 + 0.0363x	0.0173 + 0.1027x	0.00153 + 0.1365x	0.00433 + 0.3863x
TFO Mass Loss	AASHTO T 179	$x \leq 0.4$ % Loss: 0.014	$x \leq 0.4$ % Loss: 0.04	$x \leq 0.4$ % Loss: 0.055	$x \leq 0.4$ % Loss: 0.016
Penetration	ASTM D 5	$x < 60$:0.8	$x < 60$:2.3	$x < 60$:2.5	$x < 60$:7.1
		$x > 60$: 0.8 + 0.03* (x-60)	$x > 60$: 2.3 + 0.08* (x-60)	$x > 60$: 2.5 + 0.05* (x-60)	$x > 60$: 7.1 + 0.14* (x-60)
Ductility	ASTM D 113	a	a	a	a
Softening Point °C	ASTM D 36	Water or Glycerin 0.41(°C)	Water or Glycerin 1.2(°C)	Water or Glycerin 0.7(°C)	Water or Glycerin 2.0(°C)
		Ethylene glycol 0.72(°C)	Ethylene glycol 2.0(°C)	Ethylene glycol 1.08(°C)	Ethylene glycol 3.0(°C)
Solubility	AASHTO T 44	0.035	0.1	0.09	0.26

[a] Precision is presented in graphical form in ASTM D 113. It is a linear function based on the ductility result.

Dispute resolution

Disputes between suppliers (producers) and purchasers (users) of asphalt binder most commonly arise when test results measured at the purchaser's laboratory do not meet specification criteria. As discussed earlier, the variability in test results can be caused by many factors including sample handling, sample storage, operator error, and testing equipment variability. These factors add to an overall testing error, making it difficult to determine the true or actual value of any material property. Considering this, it then becomes more important to monitor the precision of test results than to say one result is correct, while another result is incorrect.

Consider the example below in which the supplier and purchaser have performed DSR testing on an unaged split sample of a PG 58-28 asphalt binder at 58°C.

Example 2
Supplier $G^*/\sin \delta$ = 1.10 kiloPascal
Purchaser $G^*/\sin \delta$ = 0.96 kiloPascal

The average value of $G^*/\sin \delta$ is 1.03 kiloPascal.

Multiplying the average value by the appropriate d2s percent (from Table 9.2 the multi-laboratory d2s percent value for DSR testing on an unaged asphalt binder is 17.0 percent) yields the following:

$$d2s = 1.03 \text{ kiloPascal} \times 17.0 \text{ percent} = 0.17 \text{ kiloPascal}$$

The difference between the Supplier and Purchaser test values is:

$$\text{Difference} = 1.10 - 0.96 = 0.14 \text{ kiloPascal}$$

Since the difference in the test values (0.14 kiloPascal) is less than the allowable d2s (0.17 kiloPascal), the two test results are considered within acceptable limits of variability. Therefore, both test results are considered admissible (or equally "correct").

The preceding example is a typical example of a dispute in test results between a supplier and a purchaser. The supplier's lab tested the sample and found it to pass the specification criterion at 58°C. The purchaser's lab also tested the sample at 58°C but found it to fail the specification criterion. Because the test results are within the acceptable range of testing variability for two laboratories, the involved parties need to decide which value is "correct". Complicating this process is the fact that inherent variability in the test procedure may make it virtually impossible to ever determine the "true" value for the measured property.

One method for resolving testing differences is described in ASTM D 3244, *Standard Practice for Utilization of Test Data to Determine Conformance with Specification*. This standard provides a procedure for resolving testing disputes between laboratories. The following steps illustrate the process:

Step 1

Determine the reproducibility (R) for the test method at the specification level using available d2s percent values from the Precision and Bias statement for multi-laboratory precision and the equation below.

$R = d2s \text{ percent} \times S$

where S = Specification criterion value

As an example, consider the test results from DSR testing on PAV-aged asphalt binder. For $G^*\sin \delta$,

d2s percent = 40.2 percent
S = 5000 kiloPascal

Therefore,

$R = 40.2 \text{ percent} \times 5000 \text{ kiloPascal} = 2010 \text{ kiloPascal}$

Step 2

Determine if the two disputed test results are within the reproducibility (R) of the test method.

As an example, suppose:

Supplier $G^*\sin \delta$ = 4725 kiloPascal
Purchaser $G^*\sin \delta$ = 5890 kiloPascal

The difference between the Supplier and Purchaser test values is:

$\text{Difference} = 5890 - 4725 = 1165 \text{ kiloPascal}$

Since the difference in the test values (1165 kiloPascal) is less than the allowable R (2010 kiloPascal), the two test results are within the reproducibility (R) of the test method. Therefore, both test results are considered admissible (or equally "correct").

Step 3

If the two disputed test results are within the reproducibility (R) of the test method, then calculate the Assigned Test Value (ATV) as the average of the two test results.

Using the data above, the ATV is determined to be 5308 kiloPascal. Even though the test results were within the allowed variability of the test, the ATV exceeds the 5000 kiloPascal specification value and is therefore considered a failing result.

If the difference between the two test results is greater than the reproducibility of the test, then a new split sample should be obtained and the tests rerun by both parties. If the second set of test results is still outside of the allowed variability, then a third test should be run by a referee lab.

The reproducibility should be multiplied by 1.2 (to convert from a range for two labs to a range for three labs), and the difference between the highest and lowest of the three values evaluated to see if it falls within the new reproducibility limits. If the difference is acceptable, then the ATV for this set of test results will be the average of the three results. If the difference exceeds the allowed reproducibility, the ATV should be assigned as the average of the two closest results. This is best illustrated by example.

Example 3

A split sample of a PG 76-22 asphalt binder is tested by two labs and the BBR *m-value* at −12°C was determined to be:

Supplier BBR m(60) = 0.323
Purchaser BBR m(60) = 0.298

The d2s percent for BBR *m-value* is 6.8 percent.
m(60) = 0.300

Therefore,
R = 6.8 percent × 0.300 = 0.020

The difference between the Supplier and Purchaser test values is:

$$\text{Difference} = 0.323 - 0.298 = 0.025$$

Since the difference in the test values (0.025) is more than the allowable R (0.020), the two test results are not within the reproducibility (R) of the test method. Based on this analysis, the two labs decide to conduct a retest on the split sample.

Retest values are determined by both labs as follows:
Supplier BBR m(60) = 0.325
Purchaser BBR m(60) = 0.296

The difference between the Supplier and Purchaser test values is:

$$\text{Difference} = 0.325 - 0.296 = 0.029$$

Once again, the difference in the test values (0.029) is more than the allowable R (0.020), indicating that the two test results are not within the reproducibility (R) of the test method.

Since the second retest also indicated values that were not within the reproducibility (R) of the test method, the two labs decide to use a third (referee) lab to test the material, with the results shown as follows:

Referee BBR m(60) = 0.302

For three labs, R is multiplied by a factor of 1.2 to give:

$$\text{BBR m(60) R} = 0.020 \times 1.2 = 0.024$$

Since the range of the three test values is still 0.029, then the data is still outside of the acceptable reproducibility for the test. As a result, the ATV is determined as the average of the two closest data values—in this case the

Referee Lab (0.302) and the Purchaser Lab (0.296). Thus, the ATV is 0.299, which fails the specification criterion.

If the Supplier BBR m(60) had been 0.320 on the second retest, then the Referee Lab would still have been required. (This is because the difference between the Supplier and Purchaser Labs would be 0.024, which is greater than 0.020.) However, by adding the Referee Lab, the R increases to 0.024 and the range of values for the three labs now would meet acceptable reproducibility. In that case, the ATV is determined as the average of the three test results:

$$\text{ATV} = \frac{0.320 + 0.296 + 0.302}{3} = 0.306$$

In this case, the ATV meets the specification criterion for the BBR *m-value* by using the average of the three lab test results for BBR m(60).

The procedure described in ASTM D 3244 provides a logical approach to solving disputes between suppliers and receivers based on the reproducibility of the test procedure. It is important for all parties involved in specification testing to agree how to settle disputes before they occur.

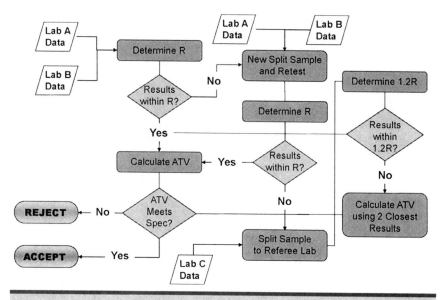

Figure 9.1 Dispute resolution flow chart

CHAPTER 10

Supplemental topics

- **Use of solubility in the specifications**
- **Asphalt rheology: mastercurves**
- **Mixing and compaction temperatures for asphalt binders**
  Mixing temperatures
  HMA and WMA
  Determining laboratory mixing and compaction temperatures
- **Temperature-Volume relationships**
  Density and specific gravity
  Temperature-volume relationship and calculations
  Specific gravity calculations
  Tank measurements
- **Commonly-asked miscellaneous questions**

Use of solubility in the specifications

According to Halstead and Welborn (1974), the first specification for asphalt in the United States was based on the use of Trinidad Lake asphalt and analytical tests to determine the amount of bitumen (portion soluble in carbon disulfide, CS_2), insoluble organic matter, and inorganic matter in a given sample. The intent of these specifications was strictly to identify the source of the asphalt. As asphalts began to be used from more than one source, specifications were written to attempt to control the quality of the asphalt. Petroleum asphalts, which came into use around 1900, were considered by some asphalt suppliers and contractors to be inferior to the Trinidad and Bermudez Lake asphalts. Thus, specifications were written to minimize the possibility of using inferior asphalt.

In 1903, A.W. Dow presented a paper describing the use of various tests, including solubility, in a specification for ensuring the quality of asphalts. The solubility test was one of the first tests used to identify asphalt and provide assurance of uniformity. The original test procedure, ASTM D 165, was adopted by ASTM in 1927 and used in the first ASTM specification for asphalt cements, D 946, published in 1947. The solubility test procedure remains substantially the same today (now identified as ASTM D 2042, adopted in 1966), although the use of solvents has changed.

The solubility test for petroleum asphalt binders, as it exists today in asphalt binder specifications, is used to ensure that an asphalt binder does not contain insoluble organic and inorganic matter that doesn't contribute to the "active cementing constituents" of the asphalt binder. In recent years, coinciding with an increased use of polymer-modified asphalt binders, a different solubility test (ASTM D 5546) was developed. ASTM D 5546 was developed principally because some polymer modifiers did not dissolve adequately during the ASTM D 2042 procedure, thus resulting in some polymer-modified asphalt binders failing the solubility criterion (usually 99.0 percent soluble or greater).

Although the original purpose of the solubility test was to identify the source and quality of native asphalts, it remains an important specification test for petroleum asphalt binders. The inclusion of a solubility requirement prevents the use of inorganic (or insoluble organic) matter that serves only to stiffen the asphalt binder, but provides no other performance benefits.

While the solubility test and criterion is in the specifications of all liquid asphalt materials, a user-producer survey conducted by the Asphalt Institute in 2003 seems to indicate that the majority of respondents (64 percent) did not require the solubility test for performance-graded (PG) asphalt binders. This may be an oversight by some users since the criterion is not included in Table 1 of AASHTO M 320, but rather is written in Section 5.4.

Given the results of the user-producer survey, the Technical Advisory Committee of the Asphalt Institute believed it was important to reemphasize that the solubility requirement should continue to be used in the specifications for all liquid asphalt paving materials including modified and unmodified asphalt cements (viscosity graded, penetration graded, and performance graded) and asphalt emulsions. This was to prevent the intentional or unintentional use of inorganic (or insoluble organic) matter in these materials. Since AASHTO M 320 includes asphalt binders that are polymer-modified, it is recommended that ASTM D 5546 is the preferred procedure for determining solubility of PG asphalt binders.

While ASTM D 5546 is the recommended procedure, it should be noted that ASTM D 2042 is also an acceptable procedure for determining the solubility of PG asphalt binders. If an asphalt binder meets the solubility requirement following ASTM D 2042, then it is not necessary to perform testing following ASTM D 5546. However, if the minimum solubility requirement is not met using ASTM D 2042, then the user should consider conducting the solubility test using ASTM D 5546 before reporting the test result as failing.

User agencies and producers can decide the required frequency of solubility testing. However, it is recommended that the solubility test is not needed on every sample, but can be performed on some scheduled basis in accordance with the producer's Approved Supplier Certification (ASC) as described in AASHTO R 26.

Asphalt rheology: mastercurves

A mastercurve, with the associated shift factors used to construct it, can be used to fundamentally characterize the behavior of an asphalt binder in which time and temperature dependency are separated. It represents the response of an asphalt binder at a specific temperature across a wide range of frequencies or time. A mastercurve can help asphalt technologists understand the rheology of an asphalt material.

Typically, a dynamic shear rheometer (DSR) is used to collect data with which to build a mastercurve for an asphalt binder, but other testing procedures and equipment may also be used. The DSR is used to test a sample at a range of temperatures and frequencies. It is generally recommended that testing is conducted at several temperatures surrounding the "reference temperature"—the temperature at which the mastercurve will be constructed. At each test temperature, dynamic modulus testing is conducted at a range of frequencies. For practical testing purposes, frequencies less than 0.1 rad/s (0.0159 Hz) are sometimes not used, nor are frequencies greater than 100 rad/s (15.9 Hz). Because data at lower and higher temperatures will be shifted to the reference temperature, some overlap in test data is preferred to produce a smooth mastercurve. This means that test temperatures should probably not be more than about 10°C apart.

Another consideration in selecting test temperatures is the need to reset the zero gap if the temperature varies too much. According to the DSR test procedure (AASHTO T 315), the zero gap setting does not have to be changed as long as the test temperature is within 12°C of the temperature at which the zero gap was set. It should be noted that some rheometers are designed so that the zero gap is maintained with changes in temperature. For these DSRs, the 12°C requirement does not apply.

In addition to temperature selection, the asphalt technologist needs to select a proper strain level for testing. If too high of a strain level is selected, then results may fall outside of the linear viscoelastic range of the binder, resulting in poor curve fitting. Similarly, too low of a strain may result in too little movement in the asphalt binder to be accurately measured.

Once testing is completed, then the analysis can begin. Figure 10.1 shows a sample of test data at multiple loading frequencies and temperatures. Fortunately, computer programs, such as the RHEA™ software, are available that simplify the job of constructing a mastercurve.

The first step in the construction of a mastercurve is to generate the isotherms. An isotherm is simply the line that is created when the storage modulus (G') or loss modulus (G'') is plotted as a function of frequency at a single temperature. Typical results are shown in Figure 10.2.

Next, the data is shifted from multiple temperatures to the reference temperature and fit to a model for complex modulus and phase angle through a series of calculations, resulting in a smooth, continuous curve composing of the data at different temperatures and frequencies (Figure 10.3).

	5°C			15°C			25°C	
Freq. [rad/s]	G' [Pa]	G" [Pa]	Freq. [rad/s]	G' [Pa]	G" [Pa]	Freq. [rad/s]	G' [Pa]	G" [Pa]
0.1	2.19E+07	1.58E+07	0.1	4.95E+06	5.29E+06	0.1	4.80E+05	8.78E+05
0.126	2.50E+07	1.72E+07	0.126	5.73E+06	5.87E+06	0.126	5.85E+05	1.02E+06
0.158	2.78E+07	1.85E+07	0.158	6.56E+06	6.47E+06	0.158	7.05E+05	1.18E+06
0.2	3.07E+07	1.98E+07	0.2	7.46E+06	7.10E+06	0.2	8.45E+05	1.35E+06
0.251	3.37E+07	2.11E+07	0.251	8.45E+06	7.76E+06	0.251	1.01E+06	1.55E+06
0.316	3.68E+07	2.23E+07	0.316	9.53E+06	8.47E+06	0.316	1.20E+06	1.76E+06
0.398	4.01E+07	2.37E+07	0.398	1.07E+07	9.22E+06	0.398	1.42E+06	2.00E+06
0.501	4.35E+07	2.50E+07	0.501	1.20E+07	1.00E+07	0.501	1.67E+06	2.27E+06
0.631	4.71E+07	2.63E+07	0.631	1.34E+07	1.08E+07	0.631	1.96E+06	2.56E+06
0.794	5.09E+07	2.77E+07	0.794	1.49E+07	1.17E+07	0.794	2.28E+06	2.88E+06
1	5.49E+07	2.91E+07	1	1.66E+07	1.26E+07	1	2.66E+06	3.24E+06
1.26	5.90E+07	3.06E+07	1.26	1.83E+07	1.35E+07	1.26	3.08E+06	3.62E+06
1.58	6.34E+07	3.21E+07	1.58	2.02E+07	1.45E+07	1.58	3.55E+06	4.04E+06
2	6.80E+07	3.36E+07	2	2.22E+07	1.56E+07	2	4.08E+06	4.49E+06
2.51	7.28E+07	3.52E+07	2.51	2.44E+07	1.67E+07	2.51	4.68E+06	4.97E+06
3.16	7.78E+07	3.67E+07	3.16	2.68E+07	1.78E+07	3.16	5.34E+06	5.50E+06
3.98	8.30E+07	3.84E+07	3.98	2.92E+07	1.90E+07	3.98	6.07E+06	6.06E+06
5.01	8.84E+07	4.00E+07	5.01	3.19E+07	2.02E+07	5.01	6.89E+06	6.67E+06
6.31	9.40E+07	4.16E+07	6.31	3.47E+07	2.14E+07	6.31	7.78E+06	7.32E+06
7.94	9.99E+07	4.33E+07	7.94	3.77E+07	2.28E+07	7.94	8.77E+06	8.01E+06
10	1.06E+08	4.50E+07	10	4.09E+07	2.41E+07	10	9.85E+06	8.75E+06
12.6	1.12E+08	4.67E+07	12.6	4.42E+07	2.55E+07	12.6	1.10E+07	9.54E+06
15.8	1.19E+08	4.84E+07	15.8	4.78E+07	2.70E+07	15.8	1.23E+07	1.04E+07
20	1.26E+08	5.02E+07	20	5.15E+07	2.85E+07	20	1.37E+07	1.13E+07
25.1	1.33E+08	5.19E+07	25.1	5.54E+07	3.00E+07	25.1	1.52E+07	1.22E+07
31.6	1.41E+08	5.36E+07	31.6	5.95E+07	3.16E+07	31.6	1.69E+07	1.32E+07
39.8	1.49E+08	5.54E+07	39.8	6.38E+07	3.33E+07	39.8	1.86E+07	1.42E+07
50.1	1.57E+08	5.71E+07	50.1	6.83E+07	3.49E+07	50.1	2.05E+07	1.53E+07
63.1	1.65E+08	5.88E+07	63.1	7.29E+07	3.66E+07	63.1	2.26E+07	1.65E+07
79.4	1.74E+08	6.05E+07	79.4	7.78E+07	3.83E+07	79.4	2.48E+07	1.77E+07
100	1.83E+08	6.22E+07	100	8.27E+07	4.00E+07	100	2.71E+07	1.89E+07

Figure 10.1 Mastercurve sample data

In each mastercurve there are four characteristic parameters that provide a description of the rheology of the asphalt binder. The glassy modulus (G_g) is a limiting modulus at very low temperatures (or high frequencies). For asphalt binders tested in shear, the glassy modulus is approximately 1 GPa (1000 MPa). On the other end of the curve is the steady-state viscosity, η_{ss}, which is the Newtonian viscosity at high temperatures (or low frequencies). The cross-over frequency, ω_c, is the frequency (at the reference temperature) where the phase angle is 45 degrees. It can be considered as a hardness parameter at the reference temperature. Finally, the rheological index, R, is an indicator of the rheological type and indicates the amount of delayed elastic behavior that will be exhibited by the asphalt binder. Past research has found that it is analogous to the Penetration Index (PI). These parameters are illustrated in Figure 10.4.

Mastercurves can be used for comparing the rheological properties of different asphalt binders as shown in Figure 10.5. In this figure, three asphalt binders of different grades are graphed at a reference temperature of 15°C. The curves have a similar shape, but are shifted, indicating a difference in modulus at a given frequency, but a similar response to changes in frequency.

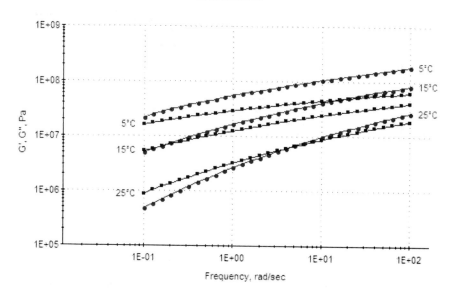

Figure 10.2 Mastercurve isotherms

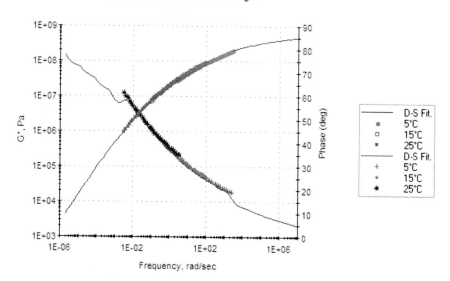

Figure 10.3 Mastercurve G* and phase angle

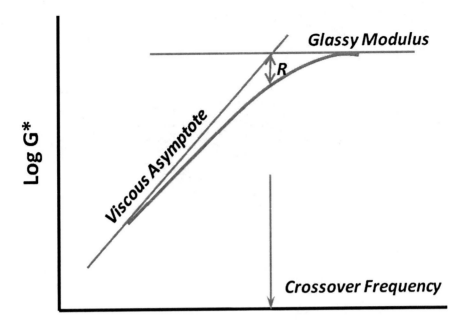

Figure 10.4 Mastercurve parameters

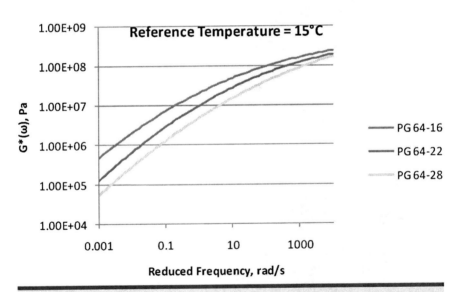

Figure 10.5 Rheological properties of different angles

Mixing and compaction temperatures for asphalt binders

Asphalt is a thermoplastic material that decreases in viscosity with increasing temperature. The relationship between temperature and viscosity, however, may not be the same for different sources or types and grades of asphalt material.

Mixing temperatures

Both asphalt and aggregate must be heated before they are mixed—the asphalt to make it fluid enough to be pumped and the aggregate to make it dry and hot enough to keep the asphalt fluid while it is coating the particles. The degree of heat in the dried aggregate controls the temperature of the asphalt-aggregate mixture because the temperature of the asphalt cement rapidly adjusts to that of the aggregate when the two are mixed.

The mixing temperature for a hot mix asphalt mixing facility is typically governed by placing and compaction requirements specific to the project. Best practice suggests that the lowest temperature that will allow time for the mixture to be hauled, placed, and sufficiently compacted should be used. Because asphalt binder properties and project conditions may vary widely, it is not appropriate to definitively recommend minimum or maximum mixing temperatures. However, to provide guidance, the Asphalt Pavement Environmental Council has published recommended ranges of storage and mixing temperatures at the hot mix asphalt mixing facility based on the grade of the asphalt binder (Table 10.1). The user is encouraged to consider these recommended temperature ranges when conducting mixing and compaction operations. The user may also wish to discuss appropriate mixing temperatures with the asphalt binder supplier. Regardless, it is strongly recommended that the maximum mixing temperature should never exceed 350°F (177°C).

Temperatures for emulsified asphalt-aggregate mixtures and cutback asphalt-aggregate mixtures are substantially lower than those required for hot mix asphalt. While these lower mixing temperatures may not provide thoroughly dry aggregates, experience has shown that they can be satisfactory for the intended purpose, especially when asphalt emulsions are used. Temperatures of 50°F–158°F (10°C–70°C) are typically used for mixing with emulsified asphalts. Typical temperatures for mixing with MC and SC cutback asphalts may range from 131°F to 239°F (55°C to 115°C) depending on the grade. RC cutback asphalts are not recommended for use in hot mixing. In any case, caution should be used when selecting mixing temperatures for cutback asphalt-aggregate mixtures in the event that the selected temperature is above the flash point of the material.

Table 10.1
Ranges of storage and mixing temperatures

Asphalt Pavement Environmental Council Best Practices
Typical Asphalt Binder Temperatures

Binder Grade	HMA Plant Asphalt Tank Storage Temperature (°F)		HMA Plant Mixing Temperature (°F)	
	Range	Midpoint	Range	Midpoint
PG 46-28	260–290	275	240–295	264
PG 46-34	260–290	275	240–295	264
PG 46-40	260–290	275	240–295	264
PG 52-28	260–295	278	240–300	270
PG 52-34	260–295	278	240–300	270
PG 52-40	260–295	278	240–300	270
PG 52-46	260–295	278	240–300	270
PG 58-22	280–305	292	260–310	285
PG 58-28	280–305	292	260–310	285
PG 58-34	280–305	292	260–310	285
PG 64-22	285–315	300	265–320	292
PG 64-28	285–315	300	265–320	292
PG 64-34	285–315	300	265–320	292
PG 67-22	295–320	308	275–325	300
PG 70-22	300–325	312	280–330	305
PG 70-28	295–320	308	275–325	300
PG 76-22	315–330	322	285–335	310
PG 76-28	310–325	318	280–330	305
PG 82-22	315–335	325	290–340	315

Use mid-point temperature for test strip construction.

HMA and WMA

It is important to note that the temperatures listed in Table 10.1 are for hot mix asphalt (HMA). Warm mix asphalt (WMA) is mixed, placed, and compacted at temperatures much lower than typical HMA mixtures. Depending on the technology used to produce the WMA, mixing and compaction temperatures could be more than 30°C (54°F) lower than the field production and construction temperatures used for HMA.

Research is currently in progress to examine appropriate mixing and compaction temperatures for WMA and the effect of these lower temperatures on laboratory mix design. The lower temperatures are expected to result in less aging of the asphalt binder compared to HMA and possibly lower absorption into the aggregate.

Determining laboratory mixing and compaction temperatures

For years, asphalt mix design procedures have used equiviscous temperature ranges for selecting laboratory mixing and compaction temperatures. The Asphalt Institute's Mix Design Methods for Asphalt Concrete and Other Hot-Mix Types (MS-2) began recommending laboratory mixing and compaction temperature ranges based on viscosity as early as 1962. At that time, viscosity ranges were specified based on Saybolt-Furol viscosity. Beginning in 1974, the Asphalt Institute switched viscosity measurements to the more fundamental unit of centistokes. In 1974, the MS-2 manual recommended viscosity ranges of 170 ± 20 centistokes for mixing temperatures and 280 ± 30 centistokes for compaction temperatures when performing a Marshall mix design. Twenty years later, the same ranges were recommended for Superpave mixture design, except that the units have been converted to metric (Pascal-seconds).

The purpose of using equiviscous mixing and compaction temperatures in laboratory mix design procedures is to normalize the effect of asphalt binder stiffness on mixture volumetric properties. In this manner, a particular asphalt mixture of the same aggregate gradation will exhibit very similar volumetric properties with a soft asphalt binder, such as an AC-10 (or PG 58-28), as with a hard asphalt binder, such as an AC-40 (or PG 70-10).

Viscosity procedure

For unmodified asphalt binders, mixing and compaction temperatures can be determined by measuring the asphalt's viscosity at a minimum of two temperatures. Historically, absolute viscosity (60°C) and kinematic viscosity (135°C) measurements were used. The two measured data points were plotted on a graph of viscosity (log-log centistokes) versus temperature (log degrees Rankine) for the particular asphalt binder. For graphing purposes, the temperature is converted from Celsius to Rankine (°R = 1.8*°C + 492). Since absolute viscosity is usually reported in poises, it is necessary to convert to kinematic viscosity using the following relationship:

$$\eta_{centistokes} @ 60°C = \eta_{poises} @ 60°C \times 100 / (0.98 \times G_b)$$

where

G_b is the specific gravity of the asphalt binder at 15°C.

In AASHTO M 320, viscosity is measured with the rotational viscometer in units of centipoise. The viscosity is then reported in SI units, Pascal-seconds (Pa-s), using the following conversion:

$$1 \text{ poise} = 0.1 \text{ Pa-s} \quad \text{or} \quad 1 \text{ centipoise} = 0.001 \text{ Pa-s}$$

$$\eta_{Pa-s} = \eta_{centipoise} \times (0.001 \text{ Pa-s} / 1 \text{ centipoise}) / (CF \times G_b)$$

where CF is a temperature-volume correction factor calculated as:

$$CF = 1.0135 - 0.0006 \, (T_{test}); \text{ with } T_{test} \text{ in } °C.$$

Assuming a binder specific gravity of 1.01 to 1.03, the product of the temperature-volume correction and the binder specific gravity varies from 0.92 to 0.96 at temperatures in the range of 135°C to 160°C. Since the product of the temperature-volume correction and the binder specific gravity (CF × G_b) is typically close to 1.0, most engineers assume:

$$\eta_{Pa\text{-}s} \approx 0.001 \times \eta_{centistokes}$$

Mixing and compaction temperatures are typically reported as ranges of approximately 5 to 7°C. In normal laboratory practice, mixing and compacting is performed at temperatures near the middle of these ranges. Therefore, the above assumption should be adequate for achieving equi-viscous temperature ranges, established as 0.17 ± 0.02 Pa-s for mixing temperatures and 0.28 ± 0.03 Pa-s for compaction temperatures. This is illustrated in Figure 10.6.

As an alternate method, the complex dynamic shear viscosity, or η^*, can also be found using the Dynamic Shear Rheometer (DSR) measurement and the following equation:

$$\eta^*_{Pa\text{-}s} = G^*/\omega$$

where

ω is the angular velocity (10 rad/s in the standard DSR procedure).

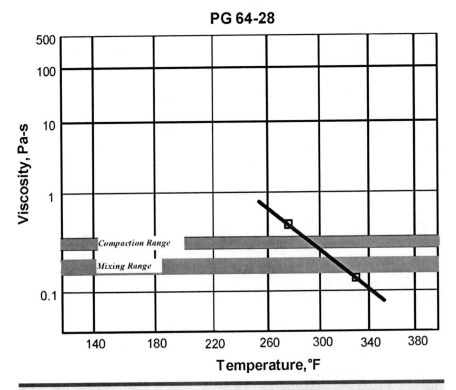

Figure 10.6 Temperature-viscosity example for mixing and compaction temperatures

By definition, the coefficient of viscosity, or simply viscosity, is actually the ratio of shear stress to rate of shear; it is a measure of the resistance of a liquid to flow. The complex shear modulus, or G*, measured in the DSR is the ratio of peak shear stress to peak shear strain. The equations for shear stress, τ, and shear strain, γ, are as follows:

$$\tau_{max} = 2 T_{max} / \pi r^3$$
$$\gamma_{max} = \theta_{max} r / h$$

where

T_{max} is the maximum applied torque (N-m),
r is the specimen radius (m),
θ_{max} is the maximum rotation angle (radians), and
h is the specimen height (m).

$G^* = (\tau_{max} - \tau_{min}) / (\gamma_{max} - \gamma_{min})$, in units of N/m², or Pa, at the peak angle in radians.

Therefore, by dividing the measured G*, in Pa, by the ω of 10 radians per second, the viscosity can be determined in Pa-s. For most practical purposes, particularly for unmodified asphalt binders, $G^* \approx G^*/\sin \delta$ at the normal high DSR operating temperatures (e.g., 58, 64, and 70°C), where δ is high (> 82°, almost purely viscous) and sin δ is about 0.99. This is a useful assumption, since the determination of $G^*/\sin \delta$ is already required for the original binder to be classified using AASHTO M 320. The results of the high temperature DSR test can then be converted directly to viscosity units by dividing by the angular velocity (10 rad/s). This conversion provides one point on the temperature viscosity graph. The second point is established from the rotational viscosity at 135°C, also required by AASHTO M 320.

For example, if the measured $G^*/\sin \delta$ of an asphalt binder is 1.212 KiloPascal at 64°C, then this is equal to 1212 Pa. To convert to viscosity units, the complex shear modulus is divided by 10 rad/s. In this case, the viscosity is calculated to be approximately 121.2 Pa-s.

The resulting graph of the viscosity-temperature relationship is usually assumed to be linear when viscosity is plotted on a double log scale (Pa-s) and temperature is plotted on a log scale (°C). This assumption can be verified by making measurements at other temperatures and adding data points to the graph. Some modified asphalt binders may not exhibit a linear trend. A blank temperature-viscosity graph is shown as Figure 10.7.

If the plot of these viscosities produces a mixing temperature higher than 175°C, it may indicate that the asphalt is modified. Because of their distinctive characteristics, modified asphalts can frequently be mixed and compacted at higher viscosities than conventional binders. While this is generally true, many modified asphalt binders may be successfully mixed at temperatures much lower than 175°C. It should be noted that temperatures above 175°C may lead to binder thermal degradation and should not be used. The binder supplier should be consulted for recommendations on appropriate mixing and compaction temperatures for modified binders for the laboratory and the field.

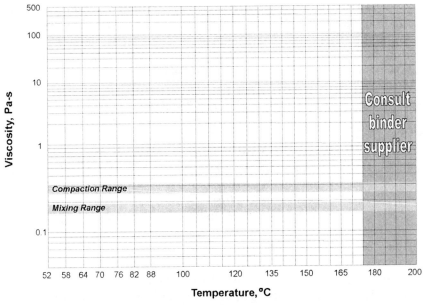

Figure 10.7 Temperature-viscosity graph

The two viscosity-based methods that may be used to develop the temperature-viscosity graph and determine mixing and compaction temperatures are:
(1) RV measurements at 135°C and 165°C, or
(2) RV measurement at 135°C and DSR measurement at the highest PG temperature where the original binder meets the specification.

The advantages of Method (1) are:
- The temperature differential between the two viscosity readings is 30°C, approximately one-half of the temperature differential of the alternate method. The assumption of linearity made when developing the temperature-viscosity graphs is more appropriate within a smaller temperature range.
- The temperatures selected for viscosity measurements, 135°C and 165°C, represent a range where most asphalt mixtures will be mixed and compacted during production.

The advantages of Method (2) are:
- The two tests (DSR at high PG temperature and RV at 135°C) do not require any additional testing.
- The temperatures for these viscosity measurements represent a range that is similar to the absolute and kinematic viscosity measurements used previously.

Phase angle procedure

Unfortunately, the increased use of highly modified asphalt binders often has resulted in the viscosity-based procedure producing "equiviscous" mixing and compaction temperatures that are unusually high (above 350°F or 177°C). At these excessively high mixing temperatures, there are potential problems with binder degradation, increased binder stiffening (aging), and excess fume generation during the mixing process. In addition, the corresponding high compaction temperatures lead to potential problems in obtaining density due to low mix stiffness, excess absorption of asphalt binder into some types of aggregates (causing a lower effective asphalt binder content), and drain-down of asphalt binder in some coarse mixes. These problems can result in high in-place air voids.

To address the problem, a recent study completed for the National Cooperative Highway Research Program (NCHRP) examined the use of mastercurves for determining the mixing and compaction temperatures for modified asphalt binders. In this procedure, a mastercurve is generated by performing a frequency sweep (0.1 to 100 rad/s) at a minimum of three test temperatures that will vary depending on the grade of the asphalt binder being tested. All asphalt binders will be tested at 80°C since it is the reference temperature in this procedure. The mastercurve is then developed (at 80°C) and the frequency is determined where the phase angle is equal to 86 degrees (as shown in Figure 10.8). This frequency (rad/s) is then input into two equations to determine mixing and compaction temperatures:

Mixing Temperature (°F) = $325\omega^{-0.0135}$
Compaction Temperature (°F) = $300\omega^{-0.012}$

The research found that the mastercurve phase angle procedure resulted in significantly lower mixing and compaction temperatures for modified asphalt binders than the viscosity procedure (Figure 10.9). The results were more reasonable and better matched field experience.

Steady Shear Flow (SSF) procedure

The Steady Shear Flow test uses high shear stress (500 Pa) in the DSR at several high temperatures to determine viscosity. An asphalt binder sample is tested in the DSR using 25-millimeter parallel plate geometry with a 500 micron (0.5-millimeter) gap. A shear stress sweep is performed at each of three test temperatures (76, 82, and 88°C) starting at 50 Pa and increasing to 500 Pa. The viscosity at 500 Pa shear stress is then plotted for each temperature and a best-fit line generated. As in the viscosity-based procedure, the data is interpolated/extrapolated to the recommended viscosity ranges to determine the temperatures for mixing and compaction. For mixing, the recommended range remains as 0.17 ± 0.02 Pa-s. For compaction, the recommended range is 0.35 ± 0.03 Pa-s.

As with the phase angle procedure, the SSF procedure results in mixing and compaction temperatures for modified asphalt binders that are

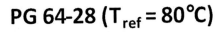

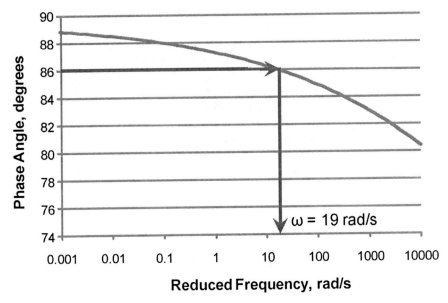

Mixing Temperature = $325 \times (19 \text{ rad/s})^{-0.0135} = 312°F$
Compaction Temperature = $300 \times (19 \text{ rad/s})^{-0.012} = 290°F$

Figure 10.8 Mixing and compaction temperature determination using phase angle procedure

considerably lower, and more rational, than the mixing and compaction temperatures determined using the viscosity procedure (Figure 10.10).

It is important to note that the temperatures determined using the procedures described above are not intended for field production. Depending on the gradation (fines content) of the mixture, the type of plant and the mixing time, the appropriate field mixing temperature for proper coating may be 10 to 30°C lower than the laboratory temperature determined by this method. Similarly, the field compaction temperature is affected by several factors: air temperature, base temperature, wind speed, haul distance, roller type, and lift thickness.

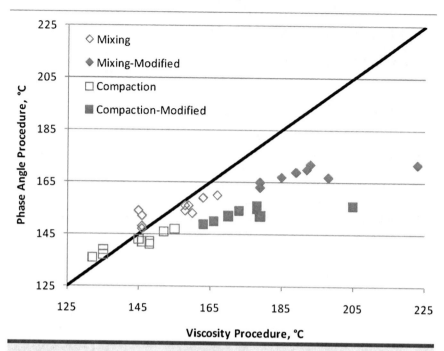

Figure 10.9 Comparison of viscosity and phase angle procedures

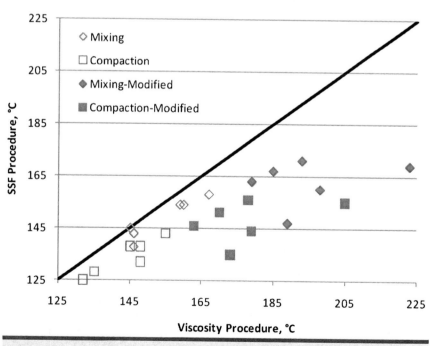

Figure 10.10 Comparison of viscosity and Steady Shear Flow (SSF) procedures

Temperature-Volume relationships

Density and specific gravity

The *density* of a material is its mass per unit volume measured at a specific temperature. It is usually given in terms of *gram/centimeter³*, *kilogram/meter³*, or *pound/feet³*. The *specific gravity* of a material is the ratio of the density of a material to the density of water, both measured at a specified temperature. It is a value that has no units. Because the density of water is 1 gram per cubic centimeter at 77°F (25°C), it is convenient to measure the density of a material at 77°F (25°C) in grams per cubic centimeter. By doing this, the specific gravity of the material is determined by dividing the density by 1.

Asphalt binder has a density, or specific gravity, very similar to water. Like water, the specific gravity of an asphalt binder changes with temperature. Additionally, different grades of asphalt binders also have different specific gravity values. Softer grades usually have a lower specific gravity than stiffer (more viscous) grades.

There is no specification requirement for the density or specific gravity of an asphalt binder because, like the temperature-viscosity relationship, it varies with different asphalt binders. However, there often is a need to know the exact density, or specific gravity, of an asphalt binder. This value is used, for example, in computing the volume of asphalt at certain temperatures for purposes of purchase payments, or in designing asphalt-aggregate mixtures.

For paving-grade asphalt binders and viscous cutback asphalts, the test is usually performed with a pycnometer bottle as previously described. For more fluid asphalt binders and asphalt emulsions, a hydrometer is used.

To indicate definite conditions applicable to a given specific gravity value, the temperature of the asphalt and water should be shown. Thus, a specific gravity at 60/60°F (15.6/15.6°C) indicates that the determination was made with both materials at a temperature of 60°F (15.6°C). The specific gravity of asphalt at 60/60°F (15.6/15.6°C) can be calculated knowing the specific gravity at 77/77°F (25/25°C). This procedure is explained in the following paragraphs.

Temperature-volume relationship and calculations

All liquids and most solids undergo changes in volume with changes in temperature. They expand when heated and contract when cooled. The change in unit volume per change in unit temperature is called coefficient of expansion, a factor that varies with variations in the density (specific gravity) of the asphalt product.

Temperature-volume correction tables for asphalt materials are published in ASTM D 4311. The multipliers given in the tables are used to convert a known volume at a given temperature to volume at 60°F (15.6°C), which is customarily used as the standard basis for volume determinations of asphalt. The equation is as follows:

$$V = V_T \times M_T \text{ (Eq. 1)}$$

where
- V = volume at 15.6°C or 60°F
- V_T = volume at given temperature
- M_T = multiplier from appropriate table

Example (U.S. customary units)

The volume of an asphalt material is measured at 180°F and determined to be 9,000 gallons. The specific gravity is known to be 0.985 at 60°F. What is the volume of the asphalt material at 60°F?

Using the data in ASTM D 4311, the correction factor, M_T, for an asphalt material with a specific gravity of 0.985 at a temperature of 180°F is 0.9587. Using the equation above:

$$V = V_T \times M_T = 9,000 \times 0.9857 = 8,628 \text{ gallons}$$

Thus, the volume of the material at 60°F is 8,628 gallons.

Specific gravity calculations

The basic formula for specific gravity of a material is the mass in air of a unit volume of the material at a specific temperature divided by the mass in air of an equal volume of water at the same temperature. This is written as follows:

$$G_x = \frac{W_x}{W_w} \text{ (Eq. 2a)} \quad \text{or} \quad G_x = \frac{W_x}{V_x * \gamma_w} \text{ (Eq. 2b)}$$

where
- G_x = specific gravity of the substance
- W_x = mass of a unit volume of the substance
- W_w = mass of a unit volume of water
- V_x = volume of the substance
- γ_w = density of water

The volume of asphalt, which changes with temperature, can be determined using the equation. However, the density of water also changes with temperature.

If the specific gravity of asphalt is found at 25/25°C (77/77°F), its specific gravity at 15.6/15.6°C (60/60°F) can be calculated by expressing *Eq. 2b* for both temperatures, combining, appropriately substituting *Eq. 1* and simplifying to become:

$$G_{15.6/15.6} = \frac{G_{25/25} * \gamma_{w25}}{M_{25} * \gamma_{w15.6}} \quad \text{or} \quad G_{60/60} = \frac{G_{77/77} * \gamma_{w77}}{M_{77} * \gamma_{w60}} \quad \text{(Eq. 3)}$$

where

γ_{w25} (γ_{w77}) = density of water at 25°C (77°F) = 0.9970 g/ml
$\gamma_{w15.6}$ (γ_{w60}) = density of water at 15.6°C (60°F) = 0.9988 g/ml
M_{25} (M_{77}) = multiplier at 25°C (77°F) from ASTM D 4311

As an example, assume the specific gravity of an asphalt cement at 77/77°F is 1.003. From ASTM D 4311, the correction factor, M_{t}, for 77°F is 0.9941. The specific gravity at 60/60°F is then:

$$G_{60/60} = \frac{G_{77/77} * \gamma_{w77}}{M_{77} * \gamma_{w60}} = \frac{1.003 * 0.9970}{0.9941 * 0.9988} = 1.007$$

Tank measurements

Many asphalt containers for transporting and storage are cylindrical tanks that are in a horizontal position. Determination of the volume of material in the tank consists of measuring its depth. Table 10.2 gives quantities in terms of percent of the total capacity based on the depth in percent of the diameter.

The volume of asphalt materials in bulk are usually expressed in gallons (liters), and the formula for determining the capacity of a cylindrical tank is shown in *Eq. 4a* for metric units (liters, meters) and *Eq. 4b* for U.S. Customary Units (gallons, feet):

$$V = 785 * D^2 * L \text{ (Eq. 4a)} \quad \text{or} \quad V = 5.88 * D^2 * L \text{ (Eq. 4b)}$$

where

V = volume (liters or gallons)
D = diameter of interior of tank (meters or feet)
L = length of interior of tank (meters or feet)

As an example, assume a horizontal tank has a capacity of 12,740 gallons and a diameter of 9 feet 6 inches (9.5 feet). If the depth of the asphalt in the tank is 7 feet 3.5 inches (7.29 feet), then the percent depth filled is:

$$\text{Percent Depth Filled} = \frac{\text{Depth}}{\text{Diameter}} * 100 = \frac{7.29}{9.5} * 100 = 76.8 \text{ percent}$$

By interpolating in Table 10.2, the volume of asphalt in the tank is

$$\text{Volume} = \frac{\text{Tank Capacity} * \text{percent Capacity}}{100}$$

$$= \frac{12740 * 82.40}{100} = 10{,}498 \text{ gallons}$$

Table 10.2
Percent capacities for various depths of cylindrical tanks in horizontal position

Percent Depth Filled	Percent Of Capacity	Percent Depth Filled	Percent Of Capacity	Percent Depth Filled	Percent Of Capacity	Percent Depth Filled	Percent Of Capacity
1	0.20	26	20.73	51	51.27	76	81.50
2	0.50	27	21.86	52	52.55	77	82.60
3	0.90	28	23.00	53	53.81	78	83.68
4	1.34	29	24.07	54	55.08	79	84.74
5	1.87	30	25.31	55	56.34	80	85.77
6	2.45	31	26.48	56	57.60	81	86.77
7	3.07	32	27.66	57	58.86	82	87.76
8	3.74	33	28.84	58	60.11	83	88.73
9	4.45	34	30.03	59	61.36	84	89.68
10	5.20	35	31.19	60	62.61	85	90.60
11	5.98	36	32.44	61	63.86	86	91.50
12	6.80	37	33.66	62	65.10	87	92.36
13	7.64	38	34.90	63	66.34	88	93.20
14	8.50	39	36.14	64	67.56	89	94.02
15	9.40	40	37.39	65	68.81	90	94.80
16	10.32	41	38.64	66	69.97	91	95.55
17	11.27	42	39.89	67	71.16	92	96.26
18	12.24	43	41.14	68	72.34	93	96.93
19	13.23	44	42.40	69	73.52	94	97.55
20	14.23	45	43.66	70	74.69	95	98.13
21	15.26	46	44.92	71	75.93	96	98.66
22	16.32	47	46.19	72	77.00	97	99.10
23	17.40	48	47.45	73	78.14	98	99.50
24	18.50	49	48.73	74	79.27	99	99.80
25	19.61	50	50.00	75	80.39		

Commonly-asked miscellaneous questions

Some material properties of asphalt binders, other than those previously described, are often the subject of questions (for various reasons). Samplings of these questions are addressed in this section.

What is the thermal conductivity of asphalt binder?

To calculate the thermal conductivity of asphalt binder, the Asphalt Institute has used the following equation:

$$K = (0.813/d) * (1-(0.0003 * (t-32)))$$

where
- d is the specific gravity at 60/60°F
- t is the temperature in F
- K is the thermal conductivity (BTU-in)/(hour-foot2-F)

This equation can be found in *Asphalts and Allied Substances: Volume IV*, by Abraham, which also has references for the equation.

What is the vapor pressure of asphalt at typical storage temperatures?

It is estimated that at a typical inventory temperature of 325°F (163°C), the vapor pressure of petroleum asphalt is less than 0.01 psia (1.5E-03 KiloPascal).

What is the typical thermal BTU value for a pound of asphalt?

The BTU value varies by temperature and the percentage of mineral matter in the asphalt. Therefore, a range is usually quoted, but an approximate value of 158,500 BTU/gallon may be considered an average value typical of an AC-10 (or PG 58) binder. Most refineries will have a calorimeter in the lab that may be used to determine this value. Actual test results should be obtained if a more accurate value is needed.

What is the typical value for the specific heat of asphalt?

The conventional method for determining specific heat for asphalt is listed as follows:

$$c = [0.388 + (0.00045 \times T)]/(d^{0.5})$$

where
- c = specific heat in BTU per pound per °F or calories per gram per °C
- d = specific gravity of the asphalt at 60/60°F
- T = temperature, °F

A typical value for specific heat for a paving-grade asphalt binder at 300°F is 0.515. This assumes a specific gravity of 1.030.

This information comes from *Asphalts and Allied Substances, 4th Edition*.

Index

A

AAG, 18
AAK, 18
AAM, 18
AASHTO T 313 – BBR, precision statement, 187
Absolute viscosity bath with tubes, 40
Absolute viscosity tests, 39–41
Accuracy, asphalt testing variability, 185–186
Aged residue (AR) system, 61
Aging effect, 19
Aging test, asphalt cements, 49–50
   RTFOT, 50`–53
   TFOT, 50
Air-blown asphalt specification tests, 159
Air-blown asphalts, 7
   overview, 151
   properties, 154
   roofing, 153
   specifications, 158–159
   tests, 154–158
AIVV tube, 40
American Society for Testing and Materials (ASTM), 3
Anionic asphalts, 122
Anti-stripping additives, 169
Antioxidants, as asphalt modifiers, 168
Apparent viscosity graph, 44
Apparent viscosity tests, 43–45
Asphalt binder grade selection, 109–114
Asphalt binder tests
   variability, 187–191
Asphalt binders, 3
   compaction temperatures, 207–213
   heating, 33–35
   HMA, 206
   laboratory mixing, 207–213
   mixing temperatures, 205–206
   sampling, 31–33
   shear rate changes, 26

   splitting, 33–35
   stiffness, 27
   thermal conductivity, 218
   WMA, 206
Asphalt cement (AC) system, 60
Asphalt cements, 3, 37
   historical tests, 37–58
   penetration tests, 45
   penetration-graded specification, 59–60
   PG asphalt binder specification, 95–118
   PG asphalt binder tests, 64–94
   phase angle, 68
   solubility test, 56–58
   specific gravity test, 58
   specifications, 59–63
   viscosity-graded specification, 60–63
Asphalt emulsions
   breaking, 128–129
   classifications, 122–124
   components, 124–128
   composition, 122
   curing, 128–129
   overview, 121–122
   production, 126–128
   types, 122–124
   uses, 121
   variables affecting quality, 129–130
Asphalt fume, 182
Asphalt modification
   additives, 169
   modifiers, 163–169
   overview, 161
Asphalt particles, in emulsion, 128
Asphalt sample configuration, 73
Asphaltic bitumen, 3
Asphalts
   air-blown, 151–159
   behavior, 25–30
   definition, 3
   emulsions, 121–129
   modification, 161–177
   natural, 4–5

petroleum, 6–9
physical properties, 23–24
production/processing methods, 7
rheology, 201–204
safety, 179–182
testing variability, 183–195
usage, 10–11
ASTM. *See* air-blown asphalt specification tests
Atmospheric distillation, 6

B

BBR. *See* bending beam rheometer
BBR mold, 79
Bending beam rheometer, 77–83
Bias, asphalt testing variability, 185–186
Bitumen, 3
BTU value, 218
Bulb, 39

C

Cationic asphalts, 122
Chemical modification, 167–168
Cleveland Open Cup (COC), flash point test, 47–49
COC flash, 49
COC manual operation, 48
COC test. *See* Cleveland Open Cup (COC), flash point test
Continuous grade, 108
Creep-recovery response, 116
Critical temperature, 107, 108
CRM. *See* crumb rubber modification
Cross-arm viscosity tube, 41
Crude selection, 8–9
Crumb rubber modification, 166–167

D

Dampproofing, 154
Deformation, PG specification, 100
Density, 214
Direct tension tester, 84–91
Dispute resolution flow chart, 195
Dispute resolution, testing variability, 187–195
Distillation products, 9
DSR. *See* dynamic shear rheometer

DSR calculations, 76
DSR creep-recovery test, 176–177
DSR operation, 69
DSR phase angle test, 174–176
DTT. *See* direct tension tester
Ductility test, asphalt cements, 53–56
Dynamic shear rheometer, 67–77

E

Elastic modification, 164–166
Elastic recovery test, 171
Elastomers, 164–166
Emulsifications, 122
Emulsifying agents, 125–126
Emulsifying plant, schematic, 127
Emulsion components
    asphalt, 124–125
    emulsifying agents, 125–126
    water, 125
ER specimens cut, 173
ER specimens recover, 173
ER test operation, 172
ER test specimen, 172
Excessive aging, PG specification, 101
Extenders, 168

F

Fatigue cracking, PG specification, 102
FD output, 175
FD test operation, 174
Force ductility test, 171–174
Functional groups
    oxidation, 17–18
    polarity, 17
Fundamental *vs.* empirical properties, 30

G

Grade bumping, 114–115
Grade classification, 106, 107
Grade verification, 105, 106

H

Handling, PG specification, 99
Historical tests, asphalt cements, 37–58
HMA. *See* hot-mix asphalt
Hot plate heating, 34

Hot-mix asphalt, 206
Hydrocarbons, 168–169

K

Kinematic viscosity bath with tubes, 42
Kinematic viscosity tests, 41–43

L

Lake asphalts, 4
Long-term stiffness, PG specification, 101
Loss on heating test, 157–158

M

Mastercurve, 201–203. *See also* rheology, asphalt
Mastercurve G*, and phase angle, 203
Mastercurve isotherms, 203
Mastercurve parameters, 204
Mixed modification, 167
Mixing and compaction temperature determination, phase angle procedure, 212
Modified asphalt binders, tests for
  centrifuge, 170
  DSR creep-recovery, 176–177
  DSR phase angle, 174–176
  elastic recovery, 171
  force ductility, 171–174
  ointment tubes, 169–170
  recovery and stress-strain tests, 170–171
  separation tests, 169–170
  solubility, 170
  toughness and tenacity, 174
Modifiers, asphalts
  antioxidants, 168
  chemical modification, 167–168
  crumb rubber modification, 166–167
  elastic modification, 164–166
  elastomers, 164–166
  extenders, 168
  hydrocarbons, 168–169
  mixed modification, 167
  oxidants, 168
  plastic modification, 166
  plastomers, 166
Mold preparation and pouring, 55

Mopping asphalts, 154
MSCR elastic behavior, 177
MSCR recovery calculation, 176
MSCR test. *See* multiple stress creep recovery test
Multiple stress creep recovery test, 115–118

N

Natural asphalt, 4–5
Non-recoverable strain, 117
Nonionic asphalts, 122

O

Ointment tubes, 169–170
Oxidants, as asphalt modifiers, 168
Oxidation, 17–18

P

PAHs. *See* polynuclear aromatic hydrocarbons
Particle size distribution, TLA composition, 5
PAV pans, 93
Pen-Vis correlation for MRL asphalt binders, 63
Penetration grading, 60
Penetration test, 157
  asphalt cements, 45
Penetration-graded specification, asphalt cements, 59–60
Penetrometer, 46
Performance graded asphalt binder tests
  aging procedure, 91–94
  bending beam rheometer, 77–83
  direct tension tester, 84–91
  dynamic shear rheometer, 67–77
  rotational viscometer, 64–67
Petroleum asphalt, 6–9
  crude selection, 8–9
  petroleum refining, 6–8
  production, 8
Petroleum refining, 6
PG asphalt binder specification, 95–118
PG binder tests
  with aging, 95
  with temperature, 98

PG specifications table, 96
Phase angle procedure, 211
Phase angle, asphalt cements, 68
Physical hardening effect, 30
Physical properties
  consistency, 23
  handling, 35
  purity, 23
  safety, 24
  storage, 35
Plastic modification, 166
Plastomers, 166
Polarity, 17
Polynuclear aromatic hydrocarbons, 19
Pouring absolute viscosity tubes, 40
Precision statement, for AASHTO T 313 – BBR, 187
Precision, asphalt testing variability, 185–186
Pumping, PG specification, 99

R

Rancho LaBrea, 5
Recovery and stress-strain tests, 170–171
Residuum content, of crude oils, 9
Rheological properties, of angles, 204
Rheology, asphalt, 201–204
Ring and ball softening point, 157
Ring-and-ball softening point test, 154–157
Rolling Thin-Film Oven Test, 50–53
Roofing, air-blown asphalts, 153
Rotational viscometer, 64–67
Rotational viscosity test, 158
RTFOT. *See* Rolling Thin-Film Oven Test
Rutting, 100
RV. *See* rotational viscometer

S

Safe handling, asphalts
  asphalt fume, 182
  storage temperatures, 181–182
Safety
  overview, 179
  PG specification, 99
  safe handling, 181–182
SBR latex modification, of hot asphalt cement, 164

SBS blending system, for asphalt terminals, 165
SBS pellets, 165
Scraping, 93
Scraping RTFO residue, into container, 54
Scraping tool, 54
Separation tests, 169–170
Shear modulus, asphalt cements, 68
SHRP. *See* Strategic Highway Research Program
Silicone mold, 71
Size exclusion chromatogram, 18
Solubility crucibles, 58
Solubility test, 170
  asphalt cements, 56–58
Solubility use, in asphalt specifications, 199–200
SP apparatus, 156
Specific gravity, 214
Specific gravity calculations, 215–216
Specific gravity pycnometers, 58
Specific gravity test, asphalt cements, 58
Specific heat, 218
SSF. *See* steady shear flow procedure
Standard emulsified asphalt grades, 124
Steady shear flow procedure, 211–213
Storage temperatures, 181–182
Strategic Highway Research Program, 17

T

T&T equipment and operation, 175
T&T output, 176
Tank measurements, 216–217
Temperature-viscosity example, for mixing and compaction temperatures, 208
Temperature-viscosity graph, 210
Temperature-volume relationships
  calculations, 214–215
  density, 214
  specific gravity, 214
  tank measurements, 216–217
Testing variability
  asphalt binder tests, 187–191
  dispute resolution, 187–195
  overview, 183
  repeatability, 185–186
  reproducibility, 185–186

TFOT. *See* Thin-Film Oven Test
Thermal conductivity, of asphalt binders, 218
Thermal cracking, PG specification, 101–104
Thermal effects, 28
Thin-Film Oven Test, 50
Time-temperature superposition, 25
TLA. *See* Trinidad Lake asphalts
TLA composition, and particle size distribution, 5
Toughness and tenacity test, 174
Trimming ductility specimen, 55
Trinidad Lake asphalts, 4–5

U

Undersealing, 154

V

Vapor pressure, storage temperatures, 218
Variability, in asphalt binder tests, 187–191
Viscosity comparison, phase angle procedure, 213
Viscosity grading, 62
Viscosity procedure, 207–210
Viscosity tests, asphalt cements
  absolute viscosity tests, 39–41
  apparent viscosity tests, 43–45
  kinematic viscosity tests, 41–43
Viscosity *vs.* SSF procedure, 213
Viscosity-graded specification, asphalt cements, 60–63

W

Warm-mix asphalt, 206
Water, 125
Waterproofing, 154
WMA. *See* warm-mix asphalt

CPSIA information can be obtained
at www.ICGtesting.com
Printed in the USA
FFOW03n0606071116
29114FF